To my son Marko who moved me to write this book.

Nenad Raos
The Cookbook of Life
(New Theories on the Origin of Life)

For the publisher
Marko Raos

Editor-in-Chief
Professor Nenad Bolf, PhD

Reviewers
Academician Mladen Juračić
Assist. Prof. Dario Hrupec, PhD
Assist. Prof. Petar Tomev Mitrikeski, PhD
David Matthew Smith, PhD

English language adviser
Marko Raos

Publisher
Provid d.o.o.
Sveti Duh 22, Zagreb, Croatia

Nenad Raos

The COOKBOOK OF LIFE

(New Theories on the Origin of Life)

Since the primordial soup was such a memorable image, I have come up with a gustatory metaphor for this exogenous delivery; please think of it as the cake mix theory.

Max Bernstein, 2006

Illustrations by Maja Raos Melis

Provid d.o.o.
Zagreb, Croatia, 2018

Nenad Raos: The Cookbook of Life (New Theories on the Origin of Life);
Published by: Provid d.o.o., Zagreb, Croatia

ISBN: 978-953-48311-1-3
Cataloguing-in-Publication data available in the Online Catalogue of the National
and University Library in Zagreb under CIP record 000989588

Technical editor: Zdenko Blažeković; Proofread: Nevenka Kopjar, PhD;
Graphic layout and illustrations: Maja Raos Melis; This edition print adaptation and
additional formatting: Marko Raos.

Publishing of this book is sponsored by:
PLIVA Hrvatska Ltd., Croatia
University of Zagreb, Faculty of Chemical Engineering and Technology
Fidelta, Ltd., Croatia

First print edition by Croatian Society of Chemical Engineers/Kemija u industriji,
Berislavićeva 6/1, Zagreb, Croatia, 2018

The question of the emergence of life on our planet is not only a biological one, for it involves chemistry, as well as geology and astrophysics. This book represents my attempt to present this complex problem to the reader using the simplest terms possible. However, as simplicity is not always synonymous with clarity, I took the liberty of employing chemical formulas and equations, as well as scientific terms, whenever I felt it necessary for better understanding of the processes involved. Their explanations are generally contained in the body of the text and are listed in the short glossary at the end of the book.

A NOTE ON CROATIAN SPELLING

Croatian spelling is strictly phonetic – a letter corresponds to one and only one phoneme, a distinct unit of sound, irrespective of its position. However, there are three notable exceptions:

nj is read as *n* in new,

lj is read as *ll* in million,

dž is read as *j* in John.

There are no letters *q*, *w*, *x*, and *y* in Croatian alphabet, but there are letters *č, ć, đ, š*, and *ž*:

č is read as *ch* in chair,

ć is read as *t* in tree,

đ is read as *dg* in bridge,

š is read as *sh* in sheep,

ž is read as *s* in pleasure.

Note also:

a is read as *a* in pat,

o is read as *o* in pot,

e is read as *e* in pet,

i is read as *i* in pit,

u is read as *u* in put,

j is read as *y* in yellow,

r is occasionally a vowel, "thrilling r" (*crkva* = church, *krv* = blood).

CONTENTS

What do cooking and theories of life's origin have in common?

FOREWORD

When I was just a young kid, our closest neighbors in Zagreb were Ivan and Marija, a married couple about as old as my parents were at the time. They had two daughters of about the same age as my twin brother and myself. Marija hailed from the northern suburbs of Zagreb, where the capital of Croatia gradually dissolves into villages and townships situated on the opposite side of the Medvednica mountain; a picturesque region called Zagorje ("over the mountain"), inhabited by peaceful yet lively and intelligent people fiercely proud of their rich vineyard country. This part of Croatia gave birth to two historic statesmen, the first president of the Communist-ruled Yugoslavia and leader of Yugoslav partisan resistance during World War II, Josip Broz Tito and Franjo Tuđman, the founder of the modern Croatian republic.

On the other hand, Marija's husband Ivan arrived to Zagreb from Dalmatia, the southern Mediterranean region of Croatia. He was born in the small town of Omiš at the mouth of Cetina River. Although the actual distance between Zagreb and Omiš is less than three hundred kilometers as the crow flies, a wild mountain range sitting right in the middle of Croatia makes all the difference; this mountain region divides Croatia into its central-European and Mediterranean halves. Inhabitants of Zagorje and Dalmatia differ from each other sharply, their languages are different, their customs are different, and – perhaps most importantly – their cuisines are very different indeed.

The original inhabitants of the city of Zagreb, *purgeri* (from German *Bürger*, citizen) prefer German or, to be more precise, Austrian cuisine. Their favorite dishes are sauerkraut with sausages or red beans while for their drink they prefer either beer or *gemišt*, a strange mixture (*Gemischt*, in German) of sour white wine and carbonated mineral water. Conversely, Dalmatians prefer sea food, especially fresh fish, olive oil and red wine, which they usually dilute with water – the ubiquitous *bevanda*, a very refreshing drink in the summer heat.

These two cuisines are, of course, incompatible and Ivan, Marija's husband, held a strong disliking for continental "specialties." He reserved particular scorn for freshwater fish and pasta prepared in German style (pasta with cucumber – yuck!). Poor Marija had to learn how to cook the Dalmatian way, spending many hours with Ivan's mother trying to imitate her way of preparing the dishes. I said "imitate" on purpose because she never had the opportunity to truly learn her mother-in-law's recipes as none of them were ever written down. Marija had only her mother-in-law's ever-changing verbal directions on how to put so-and-so or this-and-that into the pot, "as much as necessary."

To my amazement, Marija's cuisine, based on this "as much as necessary" principle turned out to be quite a pleasant surprise! I have ordered *pašticada* (a kind of rich Dalmatian meat stew) many times in Croatian restaurants, but I've never tasted as delicious a version as the one prepared by Marija. What was the secret of her (and my) favorite dish?

The recipe for *pašticada* is very simple. Take a solid portion of beef, of about two pounds, and let it sit for a few hours at slightly above room temperature (until it darkens slightly). Rub in some bacon and mediterranean herbs, roll it in wheat flour and then place it in a deep pan where you previously fried a hearty dose of onions (be careful not to drain the oil!). After a few minutes have passed, add enough red wine to completely cover the meat and then let it simmer gently. The wine needs to be sweetened though; this is traditionally done with a handful of prunes which are then cooked along with the meat. After about four hours of low-heat cooking, take out the meat, cut it in thin slices (the way it is served), and place them back into the pot to simmer for another half hour. And it's done!

And what is the secret of Marija's delicious *pašticada*? It is simply this: She used to cook beef in nothing but than pure red wine, and she never interrupted cooking until the beef began to fray on its own. This is why her dish beats the best *pašticadas* in the best Croatian restaurants, hands down. If you want to prepare a genuine *pašticada*, you must spare neither wine nor time. And this is the whole secret of Marija's favorite food.

And the same goes for the origin of life on our planet. If you want to propose a reliable theory of this marvelous event, you must spare neither geological time nor chemical transformations ("wine").

Finding the key elements that make life possible is of utmost importance here. And what elements are those? Are we talking about amino acids, nucleic bases, organic phosphates, sugars, fatty acids, or – simply – water? Is life based on a specific chemistry, i.e. proteins and nucleic acids (DNA, RNA), or is it maybe something beyond basic chemistry, such as a particular way of organizing matter? And finally, the question which transcends all other questions – what is life?

At the moment we don't even know what the very first life form looked like. All known living species have virtually the same biochemistry, the same metabolism. All living beings are comprised of cells containing double stranded molecules of DNA, composed of the same set of four nucleotides. According to the central dogma of molecular biology, a DNA sequence is transcribed into a sequence of mRNA molecules, and this mRNA sequence then forms a "mold" used for the synthesis of an amino acid sequence – a protein molecule. All proteins in each and every organism on the planet are composed out of the same set of twenty amino acids; and their sequences are synthesized, translated from the sequence of mRNA according to the same universal genetic code: a sequence of three nucleotides which encodes the position of each particular amino acid.

A multitude of different enzymes, i.e. proteins, take part in these processes of translation and transcription, as well as in the process of DNA molecule replication. A mechanism as complex as this could not have appeared spontaneously, out of nothing; it had to evolve gradually from something much, much simpler. But what did those much more fundamental organisms look like, and from which precursor entities they evolved we cannot readily deduce from biochemistries of extant organisms. We may eat and enjoy *pašticada* but we are ignorant of the exact way it came into existence.

But is there a way to reconstruct this primordial recipe? First of all, we have to find what the ingredients were. *Pašticada* is made from beef, onions, red wine, and prunes. Is this truly all we need to know?

Of course not. First we have to find a way to observe the chefs at their work. But we can only begin observing them before they start making their preparations. One of them is handling a knife, the other one is doing something to an onion, and the third chef is frantically looking for something in the refrigerator. Which one of them is going to make *pašticada*, or – perhaps a better question – which one will succeed?*

We cannot know the answers to these questions, or at least we cannot be certain of them. We do not know which theory of the origin of life is valid, which one truly leads to the first speck of living matter, the first living cell in which we can find DNA, RNA and proteins consisting of twenty amino acids. There are many chefs and many recipes – in this cookbook of life.

---

*I owe this metaphor to Max Bernstein, as told in *Phil. Trans. Roy. Soc. B* **361** (2006) 1689–1702.

Did life originate at all?

13 799
?
Big Bang

9 000
formation of
galactic disc
(beginning)

5 130

Mill. yrs from present

1. DARWIN'S SOUP

What did the first living creature look like, and how did the ancients imagine it? And what is life anyway?

It is often said that all the conditions for the first production of a living organism are now present, which could ever have been present. But if (and oh! what a big if!) we could conceive in some warm little pond, with all sorts of ammonia and phosphoric salts, light, heat, electricity, &c., present, that a protein compound was chemically formed ready to undergo still more complex changes, at the present day such matter would be instantly devoured or absorbed, which would not have been the case before living creatures were formed.

Consider a very simple recipe: take as many chemicals as possible ("all sorts of ammonia and phosphoric salts") and drop them into some hot water (" warm little pond"), add some kind of energy ("light, heat, electricity, etc...") and wait for a few hundreds of millions of years – until the first living being emerges.

And who might be our first chef, the author of this recipe? His name is Charles Darwin, a figure well known to everyone. He ventured to write this speculation ("what a big if!") in 1871, in a private letter to his friend, an English botanist named Joseph Dalton Hooker, who – as Darwin wrote in his introduction to *The Origin of Species* – "having read my sketch of 1844 – honored me by thinking it advisable to publish" and "who for the last fifteen years has aided me in every possible way by his large stores of knowledge and his excellent judgment." Though Darwin eventually did publish his *Origin* in 1859, not a drop of this "warm little pond" is to be found in this famous book.

Instead, Darwin speculated that "animals have descended from at most only four or five progenitors, and plants from an equal or lesser number."

Is there a common ancestor to all living beings? Despite his "belief" that all animals are descended from just a few distinct progenitors, and that about the same holds true for the plant kingdom, in the very next paragraph Darwin describes the very beginning of life thusly:

Analogy would lead me one step further, namely, to the belief that all animals and plants have descended from some one prototype. But analogy may be a deceitful guide. Nevertheless all living things have much in common, in their chemical composition, their germinal vesicles, their cellular structure, and their laws of growth and reproduction. We see this even in so trifling a circumstance as that the same poison often similarly affects plants and animals; or that the poison secreted by the gall-fly produces monstrous growths on the wild rose or oak-tree. Therefore I should infer from analogy that probably all the organic beings which have ever lived on this earth have descended from some one primordial form, into which life was first breathed by the Creator.

Building blocks of life

In our times this "belief" that "all the organic beings which have ever lived on this earth have descended from some one primordial form" rests on much firmer foundations. Darwin gave only anecdotal examples of the effects of poisons while modern scientists now agree that all life is based on the same basic chemistry. Today we distinguish thousands of different amino acids, and yet out of all of them only twenty are used as building blocks of all proteins – from proteins found in tobacco mosaic virus (TMV) and E. coli to proteins in a gall fly, an oak tree, not to mention those residing in our food and in our bodies. With the exception of some viruses, RNA-viruses (are viruses alive at all?) genetic information is transferred down through a myriad of generations by a single molecule, a molecule of deoxyribonucleic acid (DNA). So, what is this DNA molecule?

Surprisingly, this molecule all genes are made of is a very simple one. DNA molecules found in all living organisms contain the same set of only four nucleotides. They are composed of four nucleic bases (or nucleobases, for short): adenine (A), thymine (T), cytosine (C), and guanine (G). These bases stick helices of DNA ("double helix") together using just two kinds of Watson-Crick pairings, A-T and C-G.

And this goes even further. All protein building blocks, i.e. amino acids, are of the same chirality or handedness; they all belong to L-series. (The sole exception is glycine which cannot be chiral.) This means that amino-acid molecules cannot overlap with their mirror image. (Bodies which cannot overlap with their mirror image are called chiral – from *cheir*, Greek for hand. Molecules are chiral if they have a chiral center, in most cases a carbon atom bonded to four different substituents. This atom is usually marked with an asterisk, e.g. $NH_2C^*H(CH_3)COOH$ for the amino acid alanine.)

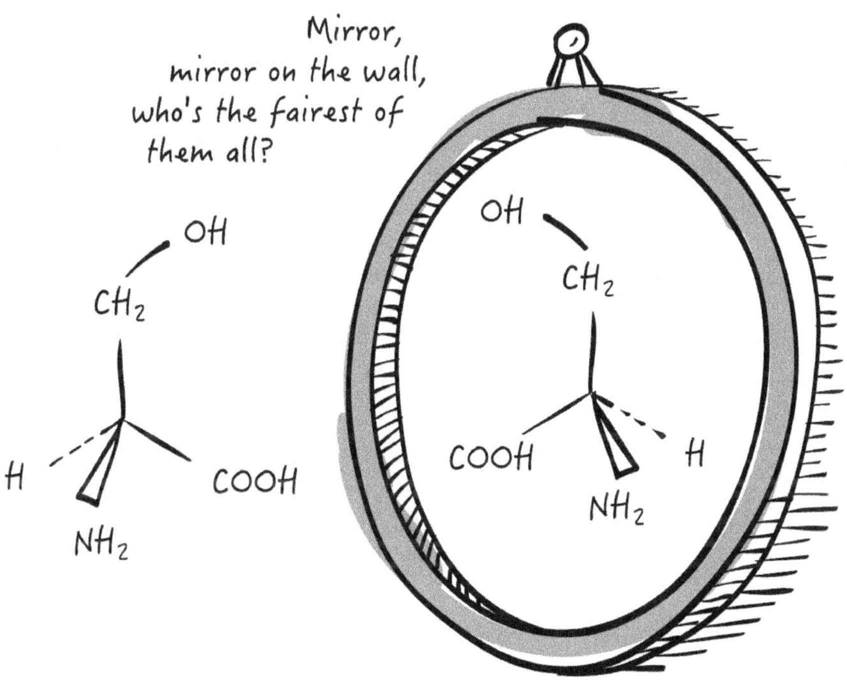

A chiral object cannot be superimposed onto its mirror image. Objects sharing this quality include biological molecules, like this one, the natural isomer (enantiomer) of amino acid serine, L-serine or (S)-2-amino-3-hydoxypropanoic acid. Chiral objects are generally asymmetrical, i.e. they lack any element of symmetry, though there are many exceptions of this rule.

Chirality is one of the principal properties of life, as present-day biochemists understand it. Naturally occurring amino acids are chiral which means that there are left and right amino-acid molecules, just as there are left and right hands or left and right feet. It is not possible to put a left hand into a right glove or a right foot into a left shoe. Enzymes which bind amino acids in order to build protein molecules recognize only "left" (L-) amino acids, just as a left glove fits a left hand only. Similarly, nucleic acids (DNA and RNA) are composed of "right" (D-) sugars (ribose in RNA, deoxyribose in DNA). Consequently, there is no living being on Earth comprised of D-proteins or L-DNA. All of this points to a single ancestor, "some one primordial form," as Darwin envisioned back in the nineteenth century.

To us, living two centuries after the great biologist, Darwin's speculations may seem to be in no way special; if there is such a thing as life then it obviously had to originate from something, from some lower form of matter, something of much lower complexity. How could it possibly be otherwise?

But the question, and – certainly – the answer, is not that simple at all. It is not that simple because we do not know what life really is. There are many definitions of life, but we lack a general one. It cannot be said, as is usually taught in schools, that all living things have three distinct properties in common – that they all feed, grow, and reproduce. While this may hold true for living creatures, crystals also eat, grow and reproduce, in their own particular ways, or course. All species evolve, but evolution is not a special property of living matter; stars evolve as well. If you state that all living beings exchange matter and energy with their environment, then you could easily come to the conclusion that your car is also a living thing.

Hunters and the hunted

The main difficulty in coming up with a definition of life is that "life" is not an actual scientific concept. (By following this line of thought to its logical conclusion it could also be claimed that biology is not a science, but this would maybe mean going a bit too far.) The concept of "life" was not invented by scientists, and not even by ancient Greek philosophers, in those hoary days before a demarcation line between science and philosophy was finally drawn. The notion of life originated at the dawn of civilization, when our ancestors roamed the earth as nomads who made their living chasing buffalo and all manner of other beasts, as Native Americans were doing right up until Darwin's time. They were hunters, and hunters are very good at distinguishing the living from the non-living.

The basis for "the hunter's definition" of life is quite simple. There are *things* which have to be hunted. They run from hunters. They move themselves. There are also *things* which have been hunted successfully. These are unable to move themselves. They can only be moved on the backs of hunters.

The first *things* are quick, alive, animated. *Things* of the second variety are motionless, dead. (As a matter of fact, the very English word "quick" originally meant "alive" as well as "fast," and similar examples can be found in most languages.) The prime task of a hunter is therefore to turn a moving, living thing into a motionless, dead one, presumably by taking something out of it. This something was called an *anima*, spirit, soul or, simply, life.

The consequence of our hunters' heritage is that, since times immemorial, life was regarded as something essentially connected with movement. The first Greek philosopher and one of the "Seven Wise Men," Thales of Miletus, "seems to have held soul to be a motive force, since he said that the magnet (i.e. mineral magnetite) has a soul in it because it moves the iron," as Aristotle reported in *On the Soul* (Book I, Part 2). Plato thought the Sun itself is alive (for no one is moving it across the sky) echoing Egyptian and Babylonian beliefs that stars are deities (which is why they exert influence on our lives, a "rationale" for astrology). In short, for Greeks all matter was in some way alive; or in Aristotle's words (in the same book): "Some say that what originates movement is both preeminently and primarily soul; believing that what is not itself moved cannot originate movement in another, they arrived at the view that soul belongs to the class of things in movement." This teaching is known as hylozoism (*hyle* = matter + *zoe* = life), and is one of the leading *leitmotifs* of Greek philosophy.

However, the fact that Greeks considered all matter as alive in some way does not mean that they assumed all beings are of the same degree of complexity. Thales, founder of the Milesian (or Ionian) school, and his famous followers Anaximander and Anaximenes, believed that there exists a simple basic substance (*arche*) from which all beings originated. This substance, which Thales thought to be water, was – as modern philosophers say – a cosmogonical rather than cosmological principle. In other words, after Thales all beings were considered not to be *composed* of, but *born* from water. Water is, in a way, the mother of all beings, living and non-living alike.

Creating flies from morning dew

Such speculations are a rational, philosophical counterpart to an ancient belief in ghosts and spirits (animism). Nevertheless, it would be unfair to say that hylozoism is just superstition, a mere belief unfounded by experience. On the contrary, hylozoism is the first philosophical notion which is a consequence of direct observation, its raw product.

As everyone can see that the Sun moves in the sky, so anyone can easily deduce that living things originate from lifeless matter. This is quite obvious in plants, which arise from earth and it is no less true for animals, especially smaller ones. "Therefore, since the most remote times, we find among various peoples all over the world the solid conviction, based frequently on observation, that the simplest animals, both of the lowest and highest order, can originate spontaneously," wrote Edmund von Lippmann in his 1933 book *Urzeugung und Lebenskraft* (Autogeneration and Vital Force). These observations eventually crystallized into a theory of spontaneous generation of life which is also known as *generatio aequivoca* (queer or enigmatic generation, in Latin).

Greek philosophers were, quite naturally, strongly influenced by this belief. While Thales taught that plants and animals all developed from amorphous slime under the influence of heat, Anaximander found a better answer to this question, claiming that every living thing arises from sea ooze and then proceeds onward in its development. For Xenophanes though, who is often regarded as the founder of Eleatic school, all organisms originate from a combination of earth and water rather than water alone.

Later philosophers developed much more sophisticated ideas. Aristotle thought that all matter has four principal properties (warm, cold, moist and dry) whose combinations create the four basic forms: water (cold and moist), earth (cold and dry), air (warm and moist) and fire (warm and dry). As for the soul, it is in various degrees present in all matter, as well as in pure, *elemental*, forms of the four elements – though earth holds less of it than water, air or fire. Consequently, earth produces plants, water produces aquatic animals and air gives birth to terrestrial animals. But what about fire? Fire, as "the lightest element" (whose "natural place" is in the heavens) produces living beings which can be found on heavenly bodies.

But not everything is that simple in Aristotle's theory. Every theory has to fit observable facts, or – as medieval scholars used to say – every contemplation has to conclude with "saving of the phenomena." To be fair to Aristotle, we are not to take his elements literally, as actual earth, water, air and fire because, as may be inferred from experience, there are many kinds of earth, water, air, and fire. Aristotle's earth, water, air and fire have to be appreciated as the *ideal* forms of matter. When the philosopher states that earth produces plants, water produces aquatic animals and air gives birth to terrestrial animals, this does not mean that those organisms are literally created from earth, water, etc. In fact, animals (such as ordinary worms, larvae of bees and wasps, fireflies and ticks) are born from morning dew, as well as from decaying slime and manure, dry wood, hair, sweat and meat.

Would you care for more examples? Flies, mosquitoes, moths, manure beetles, cantharides, fleas, lice and bed bugs are all spawned from the slimes of the seas, rivers and deep wells, as well as from the soil of the fields (humus), decaying fruit, animal excreta... There are animals which are generated from old wool and vinegar dregs. In short, there is no constraint to obtaining an animal from anything. Aristotle was quite confident in this theory, for he summed it up in a really long sentence indeed (*The History of Animals*, Book XV, Chapter I):

So also some animals are produced from animals of a similar form, the origin of others is spontaneous, and not of a similar form; from these and from plants are divided those which spring from putried matter, this is the case with many insects; other originate in the animals themselves, and from excrementitious matter in their parts; those which originate from similar animals, and have both the sexes are produced from coition, but of the class of fishes there are some neither male nor female, these belong to the same class among fishes, but to different genera, and some quite peculiar.

We now know that there are two kinds of reproduction, sexual and asexual. Sexual reproduction is achieved by fusing male and female reproductive cells called gametes or germ-cells. It is an almost universal phenomenon which is in stark contrast to asexual (vegetative) reproduction which is confined to plants and lower genera of animals, particularly insects (e.g. bees) which may hatch from unfertilized eggs (parthenogenesis). For Aristotle however, in addition to the obvious sexual reproduction, there is yet another way of generating life – a peculiar process of breeding anything from anything.

However strange and fantastic this notion may be, it is not completely groundless. It is not a product of vivid imagination but that of keen observation. Who could deny that there are multitudes of mosquitoes in the vicinity of ponds or that there are maggots in rotten meat?

And perhaps we can even create humans in this fashion! This startling claim wasn't made by Aristotle but by a scientist who lived many centuries later, a Swiss physician and alchemist named Paracelsus. (This was his nickname, "greater than Celsus" in Greek. His true name is even more curious: Philippus Aureolus Theophrastus Bombastus von Hohenheim.) He was convinced that some kind of human being, a *homunculus* (Latin diminutive of *homo*, human) could be made from "a large amount of human seed" if it was placed in a flask and left to rot in "a warm place." Newton too believed in spontaneous generation (he firmly believed in the possibility of transmutation of elements as well) and even claimed that plants were produced from attenuated emanations from the tails of the comets.

Waking up

It is often remarked that Renaissance is the "waking up" period of European history, a period when educated people and people in general, began to ask questions about "indisputable truths." From times immemorial it was believed that Earth is the center of the Universe, enclosed by seven planetary spheres (note that the Sun and the Moon are also planets, because "planet" originally meant "wandering star") and a sphere of fixed stars at the end of everything. Copernicus changed all this by placing the Sun in the center of the Universe and Giordano Bruno went on to speculate on innumerable stars with revolving planets not unlike our own.

And so the Renaissance man began to ask questions, and his first act was to open the window and simply look at the World surrounding him. *"Flieh! auf! hinaus ins weite Land!"* (Run! Go outside into the open land!), exclaimed Goethe's Faust in his famous monologue. This is why science of the Renaissance did not begin with printing of Copernicus' *De Revolutionibus Orbium Coelestium* (On the Revolutions of Heavenly Bodies) in 1543, but all the way back in 1336, with a single letter, a private report actually, by an Italian poet Francesco Petrarca on his ascent of Mount Ventoux in Provence.

He made this undertaking in the company of his brother, "each of us accompanied by a single servant." "Today I ascended the highest mountain in this region, which, not without cause, they call the Windy Peak. Nothing but the desire to see its conspicuous height was the reason for this undertaking," Petrarch declared in the first sentences of the letter addressed to Dionigi da Borgo San Sepoicro, a Franciscan monk and professor of theology. The view from the top of the mountain was marvelous:

At first I stood there almost benumbed, overwhelmed by a gale such as I had never felt before and by the unusually open and wide view. I looked around me: clouds were gathering below my feet, and Athos and Olympus grew less incredible, since I saw on a mountain of lesser fame what I had heard and read about them.

A similar kind of undertaking was described by a Croatian poet Petar Hektorović in the sixteenth century; he did not ascend a windy peak which nobody dared to ascend before, but simply went fishing with two local fishermen. They made a three-day trip in their small boat around his home island of Hvar and the two neighboring islands, Brač and Šolta, in the Adriatic Archipelago. Later on he described his "adventure" in a long poem *Ribanje i ribarsko prigovaranje* (Fishing and Fishermen's Talk), the first written record of Croatian folk songs and sayings. Strange as it may seem, but medieval noblemen who lived in their castles and palaces, or monks who spent their whole lifetimes in monasteries, hardly gave a thought to the weite Land surrounding them, just as Romeo declared in a fit of anguish (*Romeo and Juliet*, Act III, Scene III):

There is no world without Verona walls,
But purgatory, torture, hell itself.

(Such confinement to someone's native town seems almost unbelievable nowadays, but in my youth I heard a story about an old lady, still alive in the twentieth century, who spent her whole lifetime in one small village on Brač, an island only ten to forty kilometers across. And yet she still yearned to fulfil her great wish to see the sea for the first time!)

And so it came to pass that scientists began to take their first field trips, as Goethe's Faust did, "*auf! hinaus ins weite Land.*" Georg Bauer, a German physician who also made an effort to "translate" his name, as was fashionable in those days, into Latin (Georgius Agricola), made himself busy during his practice in miners' center Joachimsthal, Bohemia, by visiting mines to study the trade of miners and smelters.

His *magnum opus* was published in 1556; *De Re Metallica* (On Metallurgy), a compendium of mining and metalworking technologies and techniques. It was written in such a "modern" way that no book for miners surpassed it for two centuries.

His colleague, a physician as well as poet Franceso Redi, did not venture quite as far as that, but his eyes were keen enough to see what people around him were doing. He noticed that hunters wrap meat in cloth to prevent the appearance of maggots. He decided to put their practice to test and duplicated their actions in controlled conditions. In 1668 Redi published his treatise *Esperienze intorno alla generazione degl' insetti* (Experiments on the Generation of Insects) minutely describing the experiments which contradicted the "obvious" theory of spontaneous generation.

First, he placed meat in eight large vessels. He then sealed four of them and left the other four open to air. He soon noticed that although the meat in closed vessels did become putrid and developed a stench, it did not contain any maggots. Redi then repeated the experiment but instead of sealing the vessels, he covered them with gauze. Even though the maggots again failed to develop, the Italian scientist noticed that some flies deposited their eggs on the gauze itself. Using these observations Redi concluded, quite rightly, that decaying matter merely offers a gestating environment for emergent insects and that without fly eggs the maggots would never appear.

Even when presented with such clear evidence against the theory of spontaneous generation, Redi still did not venture to oppose it. He concluded that his experiment explained reproduction of insects only, as suggested by the title of his treatise. If maggots do develop from flies' eggs, and not from the meat itself, it does not necessarily follow that all other worms and maggots develop in the same fashion, from some kind of eggs. He still continued to believe in spontaneous generation through decay of matter; he thought it to be perfectly applicable in the instances of intestinal and wood worms or oak galls (which were allegedly generated from plant juices).

For the next two hundred years, right up until the middle of the nineteenth century, the amount of evidence against the theory of *generatio aequivoca* increased steadily, just as persistent and stubborn belief in it grew ever stronger. While no serious scientist believed that mice were generated from rotten wheat anymore, it was still quite difficult to find a scientist who would entirely rule out the possibility that life could somehow emerge from inanimate matter. Even a noted predecessor of Darwin's, a French naturalist named Lamarck, who holds the distinction of being the author of the first theory of evolutionary development, firmly believed that some primitive organisms, such as mushrooms and certain parasites, could still be created "spontaneously."

Where do maggots come from?

This Renaissance man of ours did not only move his physical body out of his castle, palace or town, but also, in effect, he managed to displace his eyes out of his body. By using his telescope Galileo discovered not only the rings of Saturn and mountains and valleys of the Moon, but he also made use of his microscope to uncover fleas, "as big as sheep." A Dutch investigator named Leeuwenhoek discovered, with the aid of a microscope of his own construction, an entirely new world of hitherto unknown "animals," *viva animalcula* (miniscule live animals). Bugs, maggots and worms in a drop of clean water? Incredible! So incredible that, as Leeuwenhoek reported, two young ladies refused to use vinegar after he had shown them the disgusting "worms" which dwelled in it!

Furthermore, when the Dutch microscopist tried dropping various decomposing substances into previously clean water, he discovered that a myriad of new *animalcula* soon appeared. He speculated that they came from the air, but his arguments were not very convincing. It is of course possible that "little animals" did come from the air, but they might as well have been generated by the putrid matter itself.

Louis Jabor, a follower of Leeuwenhoek, decided to test both hypotheses. He boiled a hay infusion for 15 minutes, and then poured equal portions into two vessels. He sealed the first vessel and exposed the second one to the air. As we would expect (with today's knowledge), germs developed in the open vessel alone. And when he later opened the sealed vessel as well, *animalcula* did eventually develop in it.

But not even these experiments managed to convince the followers of the theory of *generatio aeqivoca*. A French naturalist named Buffon proposed the existence of a vital force, a force which holds all organic matter together. He regarded all living matter as consisting of elementary living particles (presumably molecules) which do not decompose after the death of an organism, but persist until they eventually reunite into a new living being. There were many different speculations in this vein, some of them quite wild, all pointing to some unifying principle – a special force which operates in living matter alone. According to this belief, organic compounds cannot be synthesized by "inorganic" nature. And so it came as a great surprise when in 1828 a young German chemist Friedrich Wöhler synthesized "live" urea from "dead" ammonium cyanate. (We will show later that these vitalistic concepts are still an obstacle in the development of theories of life's origin.)

Jabor's experiment with boiling hay in water was repeated, improved and of course modified by numerous scientists; nonetheless, this did not place the theory of spontaneous generation into question... yet. The main obstacle was the inability to obtain, to use modern language, a sterile solution of dissolved putrefied matter. In other words, it was essential to accomplish a successful, nearly absolute sterilization of the sample, and this is not an easy task even for a modern microbiologist. (Note our persistent vulnerability to hospital infections!) The most demanding task was not sterilization of the solution itself, but sterilization of the air surrounding it. Instead of converging to a particular convincing conclusion, the experiments' results branched out into an increasing number of differing conjectures.

And thus finally, the French Academy of Sciences, as was its wont when the "immortals" expressed their wish to have some difficult and important problem solved, offered a substantial prize to any scientist who would manage to find the final answer to the question of whether it was possible to produce living matter without the aid of something already living. The prize was finally awarded to Louis Pasteur who, in 1862, published the results of his investigations. He fulfilled the requirements of the French Academy by a simple yet ingenious method. He sucked out the suspect contaminated air through a tube plugged with nitrocellulose and then dissolved this filter in a mixture of ether and alcohol. Pasteur then placed this solution under a microscope – and found a multitude of germs and other *animalcula*. This conclusively proved that it was the air that contained these organisms – and consequently that "life" did not originate from putrefied matter but from microbes dispersed in the air. (Note that this is essentially the same argument as Redi's.)

Each cell from a cell – and a helluva lot of combinations

Later on Pasteur continued to develop and improve new methods of killing various germs floating in the air. The theory of spontaneous generation lost its ground entirely, and a remark by a German pathologist Rudof Virchow, *Ominis cellula e cellula* (Each cell from a cell), written in 1860, became a fundamental and universal law of biology. Creation of living organisms out of inanimate matter, i.e. without pre-formed cells, is just not physically possible.

But then again, when in 1871 Darwin dared to write a private letter to express his belief in the origin of life "in some warm little pond," in this way he did oppose Pasteur and revitalized the discredited theory of spontaneous generation. Furthermore, he theorized that a "protein compound" is being generated on Earth right now, continually, but without the slightest chance of surviving because "at the present day such matter would be instantly devoured or absorbed." This would mean that spontaneous generation is not only possible, but that it is actually happening all the time, though invisible to us. And even more mysterious is the notion of a "protein compound." What could this possibly be? Proteins? Does this mean that proteins have an ability to form spontaneously and, also spontaneously, organize themselves into ever more complex structures, culminating in a living cell?

In Darwin's times this "protein compound" used to be a rather fuzzy concept; "protein" scarcely meant anything more precise than a mixture of nitric compounds essential for life, "albuminous" or "nitrogenous" parts of an animal's body or its diet.

Gradually, however, this vague notion did evolve. Serious study of protein structure began in 1900, led by a German chemist Emil Fischer. Furthermore, he was the first chemist who succeeded in synthesizing proteins in 1907, though his "proteinoids" were of random sequence. With this in mind let us suppose, for the sake of argument, that "in some warm little pond, with all sorts of ammonia and phosphoric salts, light, heat, electricity etc, present, that a protein compound was chemically formed" (as Darwin said) – in other words, that a molecule of natural protein did get synthesized – by pure chance.

Let us be clear on what we are talking about here, what is this "protein synthesis by pure chance?" Proteins are composed of 20 kinds of amino acids, and there are about 200 of their molecules bound together in a typical protein molecule, or a peptide chain to be exact. Two amino acids, e.g. glycine (G) and alanine (A), may form a peptide bond (–CO–NH–) in two ways; the carboxyl group (–COOH) of glycine may be bound to the amino group (H_2N–) of alanine, or *vice versa*. Therefore, from glycine and alanine two dipeptides can be formed: GA and AG.*

Let us now proceed even further. From three amino acids (let the third one be serine, S) six isomeric tripeptides can be synthesized: GAS, GSA, SGA, SAG, AGS, and ASG. For four amino acids there are 24 such combinations, for five there are 120, for six 720, and so on. We are dealing with combinations here, and he who deals with combinations routinely encounters some pretty hellish numbers! Make any assumption you wish, propose that the entire primordial ocean was filled to the brim with amino acids ready to bind together, but even then not a single molecule of any natural protein would ever get synthesized that way. Faced with such fantastic odds against this happening, a Croatian biochemist named Fran Bubanović wrote in 1917: "Therefore it is absolutely impossible that we would ever succeed in determining the exact amino acid sequence of some natural protein, even more so because we did not yet succeed to isolate – as we mentioned before – any natural protein in the pure state as a chemical individual." (In 1937 Tiselius succeeded in purifying proteins and in 1945 Sanger invented a method for determining protein sequence - never say never!)

The same line of argument is valid for the second one of Darwin's suppositions, namely that a "protein compound" would "undergo still more complex changes". Are molecules capable of self-organization? It is possible to put an enzyme into a test tube and then add a substrate in order to initiate a chemical reaction; but it is virtually impossible to create a stable system with two or more enzymes. (Such problems regularly confront scientists who attempt to develop an enzymatic method for breaking down cellulose into glucose, and ultimately into ethanol.) If this were possible, it would mean that we should be able to create living from the dead, simply by mixing a corpse's constituent materials!

I don't think that Darwin really believed this; as all organisms developed through the process of evolution, of natural selection, it is quite logical to

---

*The exact formulas of these dipeptides are $NH_2CH_2CONHCH(CH_3)COOH$ for GA (glycyl-alanine), and $NH_2CH(CH_3)CONHCH_2COOH$ for AG (alanyl-glycine).

assume that the first "protein compound", the first cell, evolved in similar fashion. But can this be proved?

Validating the Darwin's pond

An Indian guru publicly claimed that life could originate only from life, but not in a way that is readily obvious. In his view, the essence of life is the spirit (*atma*) and, as spirit is eternal, it consequently cannot be destroyed - only transformed in some way. This was too much for a certain scientist who attended the guru's lectures out of pure curiosity. He tried to oppose the guru by pointing out modern theories on the origin of life. The sage then politely asked the scientist to prove his assertions publicly, namely to create a living cell by combining chemicals in his flasks and vessels. As the scientist, also politely, declined this "challenge", the guru took this as proof positive of his assertion of the eternity of life.

But then again, there are scientists who would dare to confront the guru. In 2006 five scientists (two Americans, two Indians and a Russian) published their research results in a venerable scientific journal, *Philosophical Transactions*, circulated by the Royal Society since the eighteenth century. For its issue dedicated to the conference "Conditions for the Emergence of Life on the Early Earth" they wrote an article titled "Self-assembly processes in the prebiotic environment". In it they described an experiment involving creation of life in a test tube, or – to be more precise – in a reconstructed Darwin's "warm little pond."

Their first step was to find one such "pond" on present-day Earth. There are, of course, many ponds on this planet (though most of them are already containing such complex creatures as frogs and toads), but how do we find those which most faithfully mirror conditions on primitive Earth? The prime candidates would be the so-called volcanic aquifers, and our five scientists zeroed in on two of them, one on Kamchatka peninsula on the slopes of Mt Mutnowski, and the other one in northern California on the flank of Mt Lassen. In both pools water is held at boiling point temperature, they are free of germs, slightly acidic (pH = 3.1) and – perhaps the most important detail – they are both lined with clay.

So what did the scientists do next? First, they took a liter of water from the pool and in it they then dissolved four of the naturally occurring amino acids (glycine, alanine, valine and aspartic acid), and then they did the same with four nucleobases (adenine, cytosine, guanine and uracil), one gram each. They then added three grams of sodium phosphate, two grams of glycerol and 1.5 grams of myristic acid (one of the fatty acids). After dissolving the ingredients, they poured this solution into the boiling center of the pool (which holds a volume of about ten liters). And what happened next?

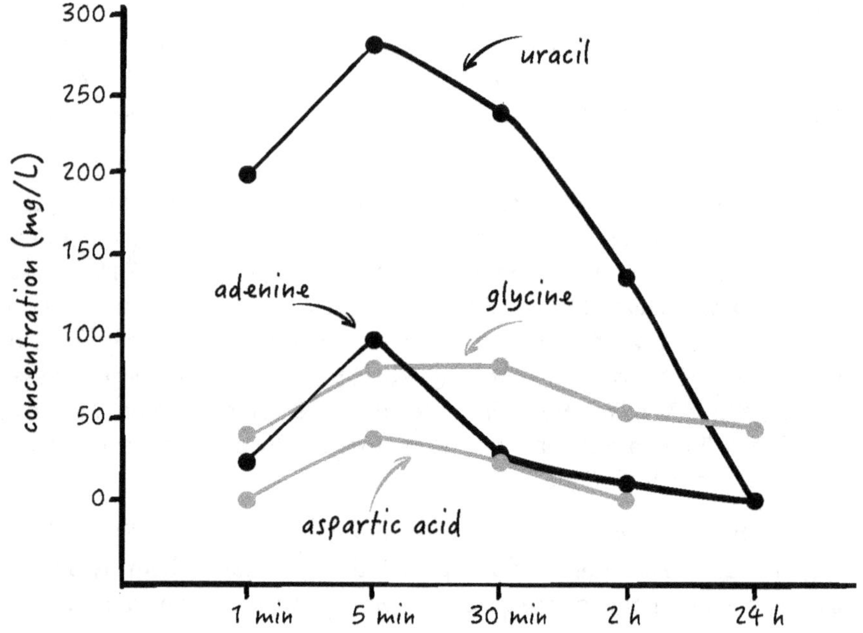

The result of a modern warm-little-pool experiment for two amino acids (glycine and aspartic acid) and two nucleic bases (adenine and uracil) at the Kamchatka site. Instead of forming vesicles (protocells), organic substances were totally adsorbed by clay minerals in two hours. Rise of uracil concentration is attributed to the deamination of cytosine, C → U. (Adapted from D. Deamer et al., *Phil. Trans. Roy. Soc. B* **361** (2006) 1809–1818.)

"After adding the organic mixture, a white precipitate appeared on the surface of the pool." What was that? You don't have to be an expert in chemistry to come to the right conclusion. Surely you already recognize this "white precipitate" from your bath tub. Myristic acid, as mentioned before, is a fatty acid, and fatty acids are constituents of soap, for soap is actually a mixture of sodium salts of fatty acids. Fatty acids, as well as soap, react with calcium and magnesium ions dissolved in water to form their respective insoluble salts – which become visible as white precipitate on water surface (and - alas - in your clogged plumbing). Something similar is exactly what happened here; the "white precipitate" was identified as insoluble aluminum and iron salt of myristic acid.

Let us proceed even further. The scientists then took 50 milliliter samples of this water following one, five, thirty, sixty and 120 minutes into the experiment; then they took additional ones 24 hours and nine days later respectively. As they took measurements of concentrations of added chemicals in all collected samples, they first noticed that concentration of cytosine is null – because it had immediately turned into uracil. And what about the other additions?

Not one of them survived. At the 120 minute mark there was no more adenine and guanine to be found, and after 24 hours there was no more uracil. And the same happened with the amino acids, all four of them. At the end of the experiment the scientists managed to detect glycerol only, alone of all the organics added to the pool. All others were adsorbed by the clay.

There is a possibility, or to be precise, there are theories that clay might be the key ingredient of Darwin's "warm little pond" because of its presumed catalytic properties, namely that clays were found capable of polymerizing amino acids and nucleotides. The scientists then tried to find some polymers in the clay, but they managed to detect only the chemicals they themselves added originally. Theoretical considerations then led them to conclude that polymerization is impossible in this instance due to low concentration of "monomers."

So what might be our conclusion at the end of this story? A hundred and forty years after Darwin, his speculations were eventually put to the test. His "warm little pond, with all sorts of ammonia and phosphoric salts, light, heat, electricity, &c,, present" produced nothing more than bath slurry. A very poor recipe. Maybe we should hire another chef to run our little kitchen of life.

Did life come from outer space?

4 600	4 590	Mill. yrs from present

beginning of the formation of the
Solar System from interstellar
cloud of gas and dust

formation of
planet embryos

2. A COSMIC BREW

Perhaps life did not originate on Earth at all but landed here from outer space. Are there microbes to be found in meteorites? And could they have survived their long voyage through interplanetary (or even interstellar) vacuum?

Let us fill up an ocean with all kinds of organic compounds, not necessarily those found in plants and animals, and wait. Wait, wait, and wait. Wait until some germ finally falls from the sky – for everything is full of germs; they are everywhere:

> *In the clearness of depth*
> *In the endlessness of seas*
> *Over stones that rivers run*
> *Or in shallows or in lakes,*

as written by the professor of microbiology Stjepan Pepeljnjak (affiliated to Zagreb University, Croatia) at the beginning of his poem *Microbes*.

Well, that's quite enough poetry for now. What about some prose?

In my youth, in the first decade after the end of World War II in Europe, we were in dearth of everything. Even the good old plain yogurt was difficult to find in stores, not to mention such exotics as pineapples and bananas (two little girls tried to eat their first bananas without peeling them, which led them to firmly decide that bananas were a rather unpalatable sort of fruit), but we were not completely deprived of that healthy beverage however, because everyone knew how to prepare their own yogurt, or – to be exact – a variety of it known as sour milk. Pour milk into a cup or clay pot, open it to the air and wait until germs from above carry out their task; in one or two days (depending on the temperature) they will happily convert fresh milk into home-made yogurt.

There are many kinds of bacteria which are capable of converting milk sugar into milk acid besides the best known one, *Bacilllus bulgaricus*

4 570	4 567	4 560	4 550
the oldest meteorite rocks	formation of the Sun		formation of Earth

("Bulgarian bacillus") which turns it into genuine Bulgarian yogurt, a drink which is said to prolong life. This sour milk of ours had a slightly different taste to that of "proper" yogurt, and was of a different consistency. Its consistency depends on how long you wait before drinking it; the protein component of milk gradually begins to coagulate and separate, and when this separation reaches maximum, you can filter it through a gauze to obtain cottage cheese. This used to be, and still is, a local delicacy in my native town of Zagreb. Each morning in the street markets you can buy fresh home-made cheese, and if you add a spoon or two of home-made sour milk cream you get *sir i vrhnje* ("cheese and sour milk cream"). Accompanied by a slice of heavy full-corn bread, bacon and onion or radish, it makes for a healthy, delicious and very refreshing breakfast. And all this is created out of nothing but milk and germs falling from the night sky.

Our second proposed recipe for creating the first life on our planet is very similar to this process of preparing sour milk, or – better yet – *sir i vrhnje*: "Fill up an ocean with all kinds of organic compounds and... wait until some germ finally falls from the sky," as I mentioned before. This theory is known as panspermia, from Greek *pan*, for all, and *sperma*, for seed. In essence, seeds of life are all around us, even in the vast emptiness of interstellar space.

We can still read heated arguments arguing for this theory nowadays, especially in popular magazines, but the theory itself is not at all new. Its birth can be traced back to the nineteenth century as a direct consequence of the collapse of the theory of spontaneous generation. If life can only originate from existing life (*Omnis cellula e cellula*, as Virchow claimed), then even the very first cell on Earth had to come from somewhere.

Pasteur's experiments really turned the theory of the origin of life upside-down: it was previously thought, during the prevalence of belief in the autogeneration of life, that life could be generated from *everything* – and now it was thought that it cannot be generated from *anything*. "Life... cannot be created on Earth; on Earth it can only sustain itself," as Croatian biologist Ivan Gjaja summed up in his 1918 panspermia theory, adding: "Life should be considered as a kind of imported goods, which is produced elsewhere."

To be fair to Pasteur, he never actually claimed anything of the kind. His experiments demonstrated only that life did not originate in our era,

4 530

core forming,
outgassing

in the conditions now present on our planet. His experiments do not, in principle, refute Darwin's "warm little pond," but they do deny "that all the conditions for the first production of a living organism are now present, which could have ever been present," as "it is often said." In 1871 however, Lord Kelvin presented a rather radical interpretation of Pasteur's experiments; he claimed that the theory of impossibility of autogeneration of life had to be accepted with the same certainty as Newton's universal law of gravitation. And so the snowball began to roll...

Panspermia and cosmozoa

The theory known as panspermia has many variants; the first one is called cosmozoa, from Greek *cosmos*, for Universe, and *zoe*, for life. The founder of this theory was a German chemist by the name of Richter. In 1865 he proposed that due to very rapid motions of heavenly bodies small particles may eventually detach from them and travel through interstellar space from planet to planet, from one planetary system to another, from galaxy to galaxy. On their journeys these particles may be accompanied by spores and other viable forms of life held in a state of suspended animation until they find a new home on some planet capable of sustaining life – just as cells of *Bacillus bulgaricus* are carried by the air until they find a cup of milk, ready to be transformed into yogurt (unless some other germs have already made their home there).

"The atmosphere of celestial bodies as well as of whirling cosmic nebulae can be regarded as a timeless sanctuary of animate forms, the eternal plantation of organic germs," as went the theory's summary by its prominent supporter Justus von Liebig, a German chemist and founder of agricultural chemistry. In other words, the problem of the origin of life transformed from a question of science to that of faith: "It is sufficient to admit that life is as old and as eternal as matter itself, and the entire argument about the origin of life loses apparently all sense by this simple admission," as Liebig explained. However, the problems persisted.

If it is not necessary to find a scientific explanation for the origin of life, some explanation for the viability of its forms still has to be offered. How is it possible for spores to survive their long voyage through emptiness of interstellar space? Which force or forces move them or, to put it in the language of physics, which force accelerates them to the proper escape

4 500

Moon-forming
giant impact

velocity, enabling them to break the chains of gravity both of their planets and their suns and thus disperse throughout the Universe? How could they survive impacts with the atmosphere, the water, and finally the hard rocks of a planet destined to be seeded with life? These germs are pioneers, pioneers of life but, in contrast to pioneers of the American West, they travel without any map or compass, and during their long travels they cannot hold the scantest notion of their whereabouts.

Helmholtz, a well-known German physicist and one of the founders of thermodynamics, believed that meteorites are the key actors in propagation of life. (This version of panspermia theory is known as lithopanspermia, from *lithos*, stone or rock in Greek.) He claimed that meteorites' interiors remain cool enough to render germs viable throughout their passage through the atmosphere. It is therefore not at all that surprising that many scientists, biologists as well as chemists, made numerous attempts to find microbes, or at least some form of organic matter, in meteorites. The greatest such "success" came to light in 1932 when Charles Lipman discovered a multitude of bacteria and their spores in the interiors of meteorites, all very similar, perhaps even identical, to terrestrial microbes.

But when, at the beginning of twentieth century, the cosmozoa theory began to lose its ground, primarily due to new developments in organic chemistry and organic synthesis particularly (to which Liebig, ironically, made a considerable contribution), the notable Swedish chemist Svante Arrhenius came out with a new theory which went along similar lines. Arrhenius is well known to every chemist as the founder of the theory of electrolytic dissociation. Not only is he considered to be one of the "three forefathers" of physical chemistry (the other two are van't Hoff and Ostwald), but he was also a scientist of very broad interests, particularly in astronomy, as well as a fluent writer and author of many popular science books. In his book *Worlds in the Making*, published in 1908, Arrhenius elaborated on his new theory, the theory of panspermia, which he had previously presented in the article *Die Verbreitung des Lebens im Weltraum* (Dispersion of life in space) in 1903. According to his hypothesis, the "seeds of life" are transported through immensities of cosmic space by means of solar wind, a natural phenomenon predicted by Maxwell and experimentally confirmed by Lebedev in 1901. (It is worth noting that pressure of solar radiation in the vicinity of Earth is very low, not more than 4×10^{-11} bar.)

4 470

large-sized impactors boiled
oceans ? times

This proposed scenario for transferring life from planet to planet is quite simple. All you need is a huge volcanic eruption, strong enough to lift life-carrying particles into the highest levels of the atmosphere – and beyond. At a high enough altitude the Sun's rays begin to push against the particles until they reach other planets, possibly even those in other planetary systems. A natural corollary to this hypothesis is that Earth, presumably, should possess a long tail, similar to that of a comet, teeming with spores of all kinds of terrestrial microbes.

And what about some concrete numbers? Let us quantify this whole notion a little bit. Assuming that spores' diameter ranges from 0.15 to 0.2 μm, Arrhenius calculated that they would traverse the border of our planetary system in 14 months and then reach the nearest star, Alpha Centauri, four light years distant, in just nine thousand years! Conversely, for particles larger than 1.5 μm across, the pressure of solar rays is not sufficient to overcome the Sun's gravity, and such particles will remain forever confined to our planetary system.

It is easy to calculate, by using nothing but basic laws of motion, the trajectories of 0.15–0.2 μm particles accelerated by the pressure of Sun's rays and its gravity. The second question however, that of survival of the bacteria and their spores, was not even touched on by Arrhenius. There is also the question of the destination stars' "solar wind" as well; this opposing wind from target stars would quite obviously decelerate incoming particles and so prevent them from reaching planets in their systems. And finally, in addition to these scientific, or to be more precise, technical problems that had to be overcome by panspermia followers, there suddenly appeared a collection of much deeper philosophical questions concerning the hypothesis of the eternity of life.

The first question is as follows: Is there any fundamental reason why life should be eternal? The complexity of life, even in its most primitive form, cannot be taken as an argument because the difference in complexity between chemistries of a "warm little pond" and the simplest bacteria is not smaller than that between bacteria and our species. Even though the theory of panspermia did originate as an answer to abandonment of the theory of spontaneous generation, both theories do have very much in common, or as Alexandr Oparin wrote:

Previously it was believed that living organisms were easily generated from

4 440

the oldest lunar rock
(anorthosite, Apollo 16)

dead matter, before our very eyes, so to speak; then the attitude was taken that life can never originate but must exist eternally. This contradiction in viewpoints is only apparent and a careful examination of the question shows that both the theory of spontaneous generation and the theory of the continuity of life are based on the same dualistic outlook on nature. Both theories start essentially with the same assumption that life is endowed with absolute autonomy determined by special principles and forces, applicable only to organisms, the nature of which is radically different from the principles and forces operative in the inanimate kingdom.

Theory of panspermia is, from a historical perspective, only an *ad hoc* answer to Pasteur's experiments. In the years following the Great War, its support by serious scientists working in the field disappeared almost completely. Panspermia was dead.

Panspermia rediviva – new theories of space germs

Mars is a faint red star in the morning sky, hundreds of millions of kilometers distant from Earth. Nobody could even dare to dream of traversing this immense empty space until, in 1903, an anonymous Russian mathematics high school teacher named Tsiolkovsky, living in Kaluga, a small town on the road to Moscow, came up with the idea of using a rocket for space travel. This crazy notion bore its first spectacular fruit soon after the end of World War II, and in 1976 the first interplanetary chunks of our planet Earth, space probes *Viking 1* and *Viking 2* found their way to the rusty red Martian desert.

And eight years later something even more fantastic happened. A Martian rock was found on Earth.

How can this be? How can it be possible to find a Martian rock on Earth? But found it was, beyond a shadow of a doubt. The rock was certainly not brought here by Martians or some other species of extraterrestrials. This rock from Mars was a meteorite ejected by an asteroid impact which struck the Red Planet and ejected rocks and debris into interplanetary space. After a period of 16 million years which it spent wandering through our cosmic neighborhood, one of those rocks found its way to Earth and fell onto Antarctic ice. There it waited for further 13 thousand years until it was found by our intrepid scientists. The diameter of this "potato-sized

4 410	4 404	4 400
the oldest terrestrial zircon (Western Australia)	formation of continental crust	

meteorite" is 15 centimeters, its mass 1.9 kilograms or 4.2 pounds, and it sports a rather unassuming name: ALH84001. This name serves as its birth certificate: it is the first meteorite (001) found in Allan Hills region (ALH) in the year 1984. (It is convenient to name meteorites according to the place where they were found – their "Earthly place of birth" so to speak.)

Surprisingly, this isn't the only Martian meteorite discovered on Earth. The first one was a four-kilogram rock found as early as 1815 near the village of Chassigny in France. In 1865 a Martian stone of similar mass was found in Shergotti, India, while in 1911 a 40-kilogram meteorite named Nakhla fell near the village of El-Nakhla in Egypt and killed a dog who happened to be in the wrong place at the wrong time. Revealing the meteorites' Martian origins proved difficult though, and the procedure for ALH84001 alone took nine years. In 1993 analyses confirmed its suspected true birth certificate, and in the summer of 1996 scientists found structures which could be interpreted as fossils of organisms hailing from the Red Planet.

It is difficult to succinctly summarize all the arguments presented in favor as well as against this interpretation. The structures found in ALH84001 appear similar to fossils of the oldest and thinnest of Earth's bacteria; some of them are egg-shaped, others are tubular. But then again, Earth bacteria are 0.5 to 20 micrometers long, and those "Martian" ones are only 20 to 100 nanometers in length – making them at least five times smaller than their terrestrial counterparts (compare 0.5 μm to 100 nm, or 0.1 μm). Such tiny organisms are considered to be too small to sustain life; note that a hemoglobin molecule is all of 5 nm across.

The second indication of the presence of life is that these structures were found in a matrix of carbonate minerals, trapped in carbonate globules which formed 3.6 billion years ago, in an aquatic environment. Despite the fact that many terrestrial organisms, such as slimes and molluscs, produce carbonates, it is quite possible that they were formed without aid from any living organism, merely by natural reaction of water and carbon dioxide with Martian minerals. This process (weathering) is well known on our planet and is responsible for formation of terra rossa, bauxite and all kinds of clays. And it is even more probable that carbonates were formed by reaction of a hot mixture (about 450 °C) of carbon dioxide and water, a vaporized soda-water so to say, immediately following the impact which launched the Martian rock. The discovery of carbonates in the Allan Hills

While *Bacillus subtilis* are 4.7 μm long (and are 0.87 μm in diameter), the size of "Martian microbes" is only 0.02 × 0.1 μm – close to the size of a hemoglobin molecule (0.05 μm). Could some kind of machinery of life be placed into it?

4 350

hydrothermal vents
dominated surface

meteorite may well represent the much-sought evidence of liquid water on Mars, and not necessarily a proof of the existence of life.

The same could be said for iron sulfides and oxide (i.e. magnetite, Fe_3O_4), which are commonly produced by anaerobic bacteria and other microorganisms here on Earth. Terrestrial bacteria regularly use magnetite as a compass, enabling them to navigate around sediments. But those earthly bacteria are much larger than the presumed Martian organisms, and then there is perhaps the most convincing argument against this idea and that is the fact that Martian magnetic field is only 0.2% the strength of Earth's, making such a mode of navigation practically useless. Moreover, there are many alternative ways to produce those minerals abiogenically.

The third and perhaps the most convincing argument in favor of the existence of life on Mars some time in its geological past is the discovery of polycyclic aromatic hydrocarbons in "Martian fossils."

"Polycyclic aromatic hydrocarbons," abbreviated as PAHs, is a long name for a bunch of very common and quite simple chemical substances. Six carbon atoms bound together in a ring and with six hydrogen atoms stuck to them constitute the well-known structure of a benzene molecule, C_6H_6. This is the molecule we will use as our starting point. By fusing two benzene rings (or benzene molecules) we arrive at the simplest of PAHs – naphthalene, a well-known moth repellent, with molecular formula of $C_{10}H_8$. By fusing three benzene rings we form two isomeric PAHs, anthracene and phenanthrene, both having the same molecular formula, $C_{14}H_{10}$, and so on. In a nutshell, all polycyclic aromatic hydrocarbons have many (*poly*) rings (*cyclic*) and these rings are all derived from a six-membered benzene ring (*aromatic*). They all consist, because they are hydrocarbons, only of hydrogen (*hydro*) and *carbon* atoms. And that would be enough chemistry for now.

PAHs are not found in living organisms, they are neither primary nor secondary metabolites – in effect many of them are carcinogens. Where did they come from, though? The answer is quite simple: every organic compound may be converted to PAHs by the simple process of heating. PAHs are actually right in the middle of pyrolytic route to graphite, which consists of bare six-membered aromatic carbon rings, without any hydrogen atoms attached. And so, the discovery of PAHs in a Martian meteorite

4 320

hydrothermal vents
dominated surface

is an undeniable proof of existence of organic compounds on the Red Planet. But can we consider this to be the proof of existence of life itself?

Not necessarily. PAHs could have been formed from Martian organics unrelated to life. There are many different PAHs found in Allan Hills meteorite, but their number is a thousand times smaller than the number found in fossils of Earth microorganisms, indicating a very simple state of organic matter on Mars. In this light, the statement "The existing standard of proof, which we think we have met, includes having an accurately dated sample that contains native microfossils, mineralogical features characteristic of life and evidence of complex organic chemistry," expressed by Dr. Zare, a member of the Martian meteorite research group, is not at all convincing. Frankly, in spite of numerous indications, none of the above mentioned "proofs" represent any real evidence for the past or present life on Mars.

But never say never; especially in science. A remote possibility that traces of Martian life have really been found in ALH84001 meteorite still exists, but this finding may be indicative of something else, something much more important. The very existence of "interplanetary" meteorites speaks in favor of the hypothesis stating that transfer of materials from planet to planet is possible, and – why not? – it would also open up the possibility that planets are capable of "exchanging" their organisms. Of course, nobody believes in the infinite perpetuity of life anymore, but life could still have originated on one planet, and then found itself transferred to another one. Perhaps life on Earth did originate on Mars and maybe one day we may even find fossils of the first terrestrial organisms there.

Saving the space germs

Perhaps life on Earth did originate on Mars... Perhaps. Communication between the two planets is possible, as proved by Martian meteorites, but can we really consider meteorites as suitable vehicles for the transfer of life? This is an entirely different question. Arrhenius and other scientists of his age speculated on this possibility, but their sole information on the nature of interplanetary space was that there was no air in it; they had no notion of the existence of radiation, much less of its deadly effects. Is it possible for bacterial spores to survive millions of years in the void of outer space, exposed to all forms of radiation and then endure the heat produced by their "rock space ship's" entrance into the Earth's atmosphere?

4 290

hydrothermal vents
dominated surface

But let us slow down a bit and take this step by step. Arrhenius speculated that volcanoes are instrumental in spreading life throughout the Universe. Scientists now know that meteorite impacts exerted much more influence on Earth's early geological history than it was previously assumed. Planetary collisions with meteorites, as well as with asteroids and planetesimals (as those small celestial bodies are called) represented key events in their formation. For example, our Moon came into existence when Earth, then only a couple of hundreds of millions years old, collided with a body of about the size of Mars. The impact of a 10-kilometer-diameter asteroid on the Yucatan peninsula caused the extinction of dinosaurs 65 million years ago marking the end of Mesozoic era.

Energy released by such collisions can be tremendous. Typical velocity of bodies entering Earth's atmosphere is 17 km/s. It is not difficult to calculate the kinetic energy of such a body: a one-kilogram meteorite has the energy of 1.45×10^8 J or 145 MJ. This may not seem much because a kilogram of gasoline or natural gas releases 50 MJ. On the other hand, this energy is released in but a fraction of a second, and there are quite huge masses involved. It was calculated that a 1.3×10^{20} kg asteroid (a 440 km object) hitting Earth with a typical velocity (17 km/s) would evaporate all our oceans and undoubtedly kill all living creatures on the planet.

But still, there are many objects regularly falling on Earth, as well as on other planets of our Solar System. Most of them burn out high in the atmosphere and are perceived as shooting stars; if they are big enough they reach the Earth's surface as meteorites or spectacularly explode in the air. The biggest ones create huge meteoritic craters accompanied by giant explosions causing earthquakes and tsunamis. Most of their energy however is released as heat – the falling space object evaporates entirely and neighboring rocks melt away. These meteoric vapors then ascend into the stratosphere where they condense into tektites – a stony rain or, to be more precise, stony snow. This snow is not white and soft though, but coal black and its glassy flakes are, typically, thumb-sized. As protection from this strange snow a woolen cap does not suffice; a Kevlar helmet would certainly be more appropriate.

Not everything involved in these events is searing hot though. The impact's shock wave breaks the bedrock and a portion of the energy is released as a vertical force which, in the spall zone, ejects rock fragments far and wide. The temperature in this zone is rather low, between 100 and

4 260
|

hydrothermal vents
dominated surface

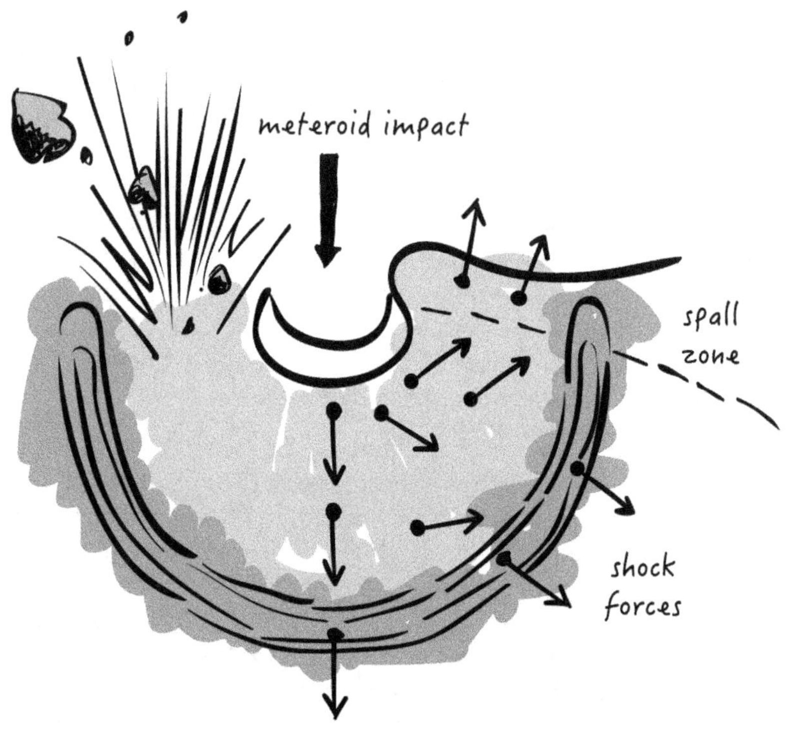

A lot can happen following an impact from space; the impactor's body and terrestrial rock partially evaporate and some of it melts; however, a portion of fragments in the spall zone may be propelled with velocities quite sufficient to reach other planets.

600 ºC or even less. It was estimated that the body of the Martian meteorite ALH84001 reached temperatures, astonishingly, no greater than 40 ºC!

It is not at all impossible for a rock fragment ejected from the spall zone to reach escape velocity, 11.2 km/s on Earth, or only 5 km/s on Mars, a planet that is only half of Earth's size and nine times less massive. This leads to an interesting possibility that life on Earth actually originated on Mars. The hypothesis is supported by findings of numerous traces of liquid water which seems to have flowed freely on the surface of the Red Planet at the time when first life appeared on Earth. "There is no consensus whether life exists, or had ever existed, on Mars, but early Mars, with then vigorous volcanism, may have been a kinder, gentler habitat than Earth," wrote E. G. Nisbet and N. H. Sleep in their February 2001 *Nature* article. "Large impacts would have oc-

4 230

hydrothermal vents
dominated surface

curred here too, but there was no so deep ocean which could be vaporized to maintain a global steam bath. Major impacts would have ejected many fragments from the surface of Mars, some of which would have landed on Earth."

There are 28 Martian meteorites, or SNCs (named according to the places – Shergotty, Nakhla, and Chassigny – where the first three of them were found) known so far, though calculations indicate that many more must have reached the Earth. Within the last four billion years, billions of Red Planet fragments were ejected into space. To be more exact, it is estimated that more than a billion fragments two-meters in diameter or larger were ejected, and that their temperature never reached 100 °C. If some kind of life on Mars did exist at some point, at least a few of its microbes should have found their way to Earth.

But what fraction of them embarked on the journey, and when did it happen? It is estimated that it takes eight million years, on average, for a Martian meteorite to reach Earth, and that only 5% of Martian meteorites were capable of accomplishing this. Therefore, about fifty million Martian rocks must have found their way to our planet. "If Mars did have life, possibly one or more cells survived impact on Mars, ejection, freezing in space, and transfer to and landing on Earth, where the cell line survived..." as claimed by authors of the *Nature* article. It is quite clear now that presumed Martian organisms would have survived the impact, but whether they would have been capable of surviving the prior eight million years-long journey through interplanetary space is a question of its own. The answer would come from research whose aim was to depict the Universe as a much more inhospitable place than it already is.

Killing the space germs

Although the English word quarantine differs from the Italian word *quarantina* in a single letter, its meaning is quite different; *quarantina* means "about forty," namely the number of days that have to be spent in isolation to be reasonably certain of preventing any possibility of epidemics. The first historical account of this institution was found in the fourteenth century's Proceedings of the Dubrovnik Senate, a council of nobles who ruled the city-republic of Dubrovnik situated on the Adriatic coast. They prescribed a 40-day period of confinement for all merchants entering *grad* (the City), as the inhabitants proudly call their native town even today.

4 200

smaller, non-sterilizing
impacts dominated

In the Middle Ages Dubrovnik was an important maritime center of trade on the Adriatic route to Levant. It is the birthplace of Croatian literature and native city of Roger Boscovich, an eighteenth century physicist and mathematician who, according to John D. Barrow (in his famous book *Theories of Everything*), "was the first to envisage, seek and propose a unified mathematical theory of all the forces of Nature." Dubrovnik is a now a tourist mecca with many hotels, restaurants and welcoming inhabitants who, regardless of their hospitality, hate to be treated as guests in their own *grad*. *Gospari* (Lords), as they are mockingly called by the rest of the country (though they use the same title to greet their guests from other regions of Croatia), have a big problem with tourist cruisers. This is not because of poor harbor facilities (which are quite adequate) but because they believe that a certain unpleasant type of sea scum, occasionally found near their beaches, has its origin in the sewage systems of those huge ships. It is amusing to lounge on a sunny beach, looking at the waves and the brilliantly clear blue sky (it is said that it has a special shade of blue, but I cannot confirm it because of my color blindness), and discuss with local people where did that gunk come from. There is a sewage system in Dubrovnik, of course, and many ships cross the Adriatic Sea...

This gunk might as well have fallen from the Moon. There are more Moon meteorites than Martian ones on Earth, of course, because Moon is much closer than Mars and its escape velocity is only 2.27 km/s, half that of Mars. But there are no suitable conditions for life on the Moon, and that is why (as an old Croatian joke from the Communist era goes) it's dead certain it is ruled by Communists.

I do not know if Americans were really that fearful of Russians and their international Communism when they decided to go to the Moon, but when its journey lasting exactly eight days, three hours, 18 minutes and 35 seconds came to end and Apollo 11 spaceship *Columbia* splashed in the Pacific Ocean on July 24, 1969, they were certainly terrified of Moon germs. Divers provided the three astronauts with Biological Isolation Garments (BIGs) and they were subsequently scrubbed down with an old-fashioned sodium hypochlorite solution. The astronauts then had to endure 21 days of isolation in a special Mobile Quarantine Facility (MQF) and God only knows why not the full forty. As nobody in a democratic country may be imprisoned without some kind of legal justification, a special Extra-Terrestrial Exposure Law had to be passed by the US Congress, just in the nick of time. Crews of the two subsequent Apollo missions, Apollo 12 and Apollo

14 (Apollo 13 did not manage to land on the Moon due to an accident) had to endure the same treatment before the quarantine process was finally dropped – it was decided that Moon rocks do not contain any trace of life.

But what about the very real possibility of microbes surviving on the Moon and even in outer space during long interplanetary voyages? Solar irradiation near Earth (solar constant) is 1.362 kW/m^2, exactly. Of this radiation, 48% is in the visible and 45% in the infrared part of the solar spectrum. Ultraviolet radiation amounts to only 7%, but this is quite enough for it to be a thousand times more lethal in space than on Earth where we are protected by the ozone layer. Is UV radiation strong enough to kill all organisms in all meteorites and thus make transfer of life from planet to planet impossible?

It is not an easy question to answer. It depends on the kind of microorganisms as well as the type of the carrying meteorite. Bacteria themselves are not very durable, but their spores are. Bacterial spores which had successfully hibernated for several millions of years have been found in amber, as well as in a 250 million year-old halite, i.e. NaCl crystal. In 1993 Gerda Horneck reported results of space experiments which have shown that 70% of bacterial and fungal spores survived a ten day exposure period to space vacuum, granted that they were additionally shielded from UV radiation. In NASA's LDEF (Long Duration Exposure Facility) mission, spores of *Bacillus subtilis* were exposed to harsh space conditions for six years. Of those protected by salt crystals a third (30%) survived. Glucose shielding proved even more advantageous. After six years in space, continually subjected to vacuum and hard radiation, 70% spores of *B. subtilis* survived if they had the advantage of being embedded in glucose crystal.

But even this proved not to be the best protection available. In the ESA BIOPLAN facility attached to Russian FOTON satellite, three types of experiments were performed. In the first one, type *B. subtilis* spores were exposed to space vacuum while shielded from UV radiation by a quartz window or a thin layer of clay. After two weeks it was found that only one spore out of a million, or even less, survived. But if they happened to be protected by being mixed with powdered clay, rock or meteorites, their survival rate increased 100,000-fold; about one in ten spores survived. Nearly total survival rate was achieved if spores were embedded in artificial meteorites.

4 140

smaller, non-sterilizing
impacts dominated

Is it then possible to transfer bacterial spores, at least, from planet to planet? It is difficult to say. An almost hundred percent survival rate in two weeks corresponds to maybe 0.01%, or even slighter, survival rate in ten million years. There are hundreds of millions or even billions of organisms in a single gram of Earth's soil, so even one to a million survival ratio may be acceptable. According to C. Mileikowsky, lithopanspermia should work only if at least a hundred spores per gram survive. According to his calculations such requirements could be met if the spores spent 300,000 years in space, shielded by 10 centimeters of meteorite rock or, conversely, one million years protected by a meter-thick shield. If they were situated a few meters below the surface of their space object, they should be "infectious" even after 25 million years. Ben C. Clark arrived to similar conclusions. He predicted survival time of 1.5 million years behind 10 cm meteorite shielding and 7.5 million years at the depth of one meter.

The eleven ifs

But this provides us with only one part of the answer. In 2001 the same scientist, Ben Clark, proposed a mathematical relation similar to the well-known Drake's (or Green Bank) equation which is used in estimating the number of technically advanced civilizations in our Galaxy. However, in contrast to Drake's equation which operates with six probability parameters (formation of planetary systems, number of habitable planets, etc.), Clark's formula evaluates 11 separate factors governing the transfer of organisms from one planetary body to another. In a few words, lithopanspermia is possible only if the incoming meteorite hits a biologically inhabited zone, if a rock from this zone is then ejected into escape orbit, if this ejected rock material survives the launch and finally if organisms thus ensconced in ejecta survive their space sojourn. Of course, positive answers to these first four *ifs* alone would still not suffice. This would only make certain that somebody somewhere could possibly find some viable spores in a pretty unusual piece of rock floating in outer space. First of all, these spores have to arrive to Earth sound alive, and then be able to propagate there.

Seven other *ifs* address these probabilities – they are the last seven factors in Clark's equation. Spores would survive if they could tolerate descent through the atmosphere of the target planet and then endure the subsequent impact with its soil (the fifth and sixth *if*). Could this be possible? Meteoroids burn in the air and smaller ones (from about one gram to one kilogram) disintegrate entirely. The big ones explode (we call them "bolides") and some even evaporate completely following their collision with Earth. But all those are of asteroid size and were surely not formed by spallting. The smallest of meteors, called cosmic dust (a few micrometers in size), do not burn at all – though such tiny particles are wholly incapable of protecting spores from cosmic radiation. In contrast, surfaces of moderately-sized meteors do melt (up to a depth of one centimeter) but their interiors still remain cool; the descent through the atmosphere is too short for heat to be conducted to any significant depth.

Though a meteorite's impact with Earth can be of tremendous force, with speeds in excess of 10 km/s, still the results of experiments reproducing such events are all but definite. Spores survived gunshots, with a survival rate of up to 90%, when they were accelerated to 4.6×10^6 m/s^2 (or $460{,}000 \times g$), but they don't manage as well in shock waves produced by controlled explosions of high explosives which can produce temperatures of up to 390 ºC for a few microseconds, and then down to 250 ºC slowly decaying over a period of several minutes. In one such experiment only 0.01% of the spores survived, just one in ten thousand.

The seventh *if* is that transfer of life would be possible only if the organism could leave the meteorite interior following the impact. Then it has to avoid being killed outright by the new environment's harsh conditions (the eighth *if*) and it has to escape the potential predators (the ninth *if*). If it manages to survive all this, it will at long last begin proliferating on its new planet, but only if there is an environment suitable for its development, that is its further evolution. And at last – the final, 11th *if* – it and its descendants have to succeed in the struggle for existence in their new biosphere.

It is not possible to deduce anything conclusive from those assumptions without knowing the exact state of both planets. Molecular oxygen is highly toxic for anaerobic organisms and some bacteria (including those found in our intestines) thrive on arsenic, a well-known poison from Agatha Christie stories. There are no toxic substances as such, just like there are no intrinsically inhospitable environments. Life is a product of its biosphere, just as the biosphere is a product of not only of its rocks and minerals, but also of all the living creatures it supports.

4 080

smaller, non-sterilizing
impacts dominated

As a result of all those unknowns, it is difficult to even approximate a reasonable value for Clark's equation. But let us suppose, for argument's sake, that probability of an event is one percent (0.01 or 10^{-2}). In this case the value of PAB (probability of transferring life from planet A to planet B) should equal $(10^{-2})^{11} = 10^{-22}$. This means that only one out of 10^{22} bacterial spores in a region of Martian soil hit by a space projectile would put roots on its new home on Earth. It is a rather small probability, but it has to be taken into consideration that there is about a hundred million (10^8) or even a billion (10^9) microorganisms in just one gram of Earth's soil (and presumably a similar amount in that of Mars). Also, following Mileikowsky's calculations, there were perhaps as much as a trillion (10^{12}) fragments ejected from Earth (and a similar number from Mars) in their long geological histories.

It is possible then that my estimate of probability factors as 10^{-2} is perhaps too conservative. But should we still unquestioningly accept statements such as the following: "Therefore, based on recent observations, experimental results and calculations, we can conclude that all the steps required for a viable transport of microorganisms between the planets of a solar system can be fulfilled by the Litho-Panspermia mechanism", as three European scientists (P. Clancy, A. Brack and G. Horneck) wrote in their book *Looking for Life, Searching the Solar System*, published in 2005? "Shall we follow the deception of the thrush," in T. S. Eliot's words, as scientific truth? There are many arguments in favor of transferability of life from planet to planet, though we still cannot claim a single undisputable proof for the notion.

At any rate, an interplanetary transfer of life would certainly not be a quiet event. Perhaps life did arrive from the sky – but this was probably less akin to brewing of sour milk, the yogurt of my childhood, but more like trying to cook a soup by throwing a bomb into a nearby pot.

4 050

Mill. yrs from present

hydrothermal vents
dominated surface

4 020
Mill. yrs from present
hydrothermal vents
dominated surface

Did life originate from some kind of broth?

3. BOUILLON À LA OPARIN

Is it possible to create new life by a simple mixing of chemicals? Many have tried to go down this route and some even claimed that they managed to produce "living forms" – but what these "living forms" were is a whole new question.

"Let us assume that in some archaic water basin on our planet coacervate droplets appeared as a result of the mixing of solutions of high-molecular organic substances," the notable Russian biologist Aleksandr Ivanovich Oparin provided us with this very brief recipe for the creation of life, the first life, of course. In other words, if you mix two thin solutions, small droplets will eventually appear – the first, ahem... *living* cells. Or, as Oparin's supporter and co-founder of the first modern theory of the origin of life, English geneticist John B. S. Haldane said, life emerged from "a hot thin soup," closely describing the condition of the Earth's first ocean.

This "hot thin soup" is, let's get to the point, another name for bouillon, the pride of French cuisine. In it one might very well find Oparin's droplets, for bouillon consists of "high-molecular organic substances" and is, in its essence, a colloid solution of proteins extracted from meat and bones.

Russians are quite fond of France, its culture and especially its language. Did you know that the first sentence of Tolstoy's War and Peace is written in French (*Eh, bien, mon prince, Gênes et Lucques ne sont plus que des apanages, des* властелинств *de la famille Buonaparte.*) as are many other paragraphs in this, perhaps the best Russian novel of all time? And so it seems quite appropriate to compare conditions on early Earth, as proposed by the Russian scientist, with some tasty French bouillon. Those similarities may go even further, assuming heterogeneity of ingredients, with Russian *borscht*, a soup prepared from cabbage and beetroot.

My mother used to perform kitchen experiments by preparing strange and unusual dishes. One of those was *borscht*, as absolutely unpalatable to me today as it was back then. It had the smell, and much of the taste, of

3 960

hydrothermal vents
dominated surface

a soup prepared from dirty dish-rags. I was subjected to this terrible smell of over-boiled cabbage once again, thirty years later, in a restaurant in Russia, one of the best hotels in Moscow (it was razed to the ground soon after – but probably not because of its *borscht*…).

There are many variants of this Russian favorite. You may take all *borscht* ingredients out of water and then add crushed nuts and some special spices (*travki*). This used to be the specialty of Katya, a piano teacher from Georgia, who used to visit Zagreb occasionally during the Soviet era. She would arrive by train and invariably get held up by the police for carrying bags of sugar, a notoriously suspicious white powder. After supper she would ask her guests, including me, how we liked it. "I wouldn't know" I replied, not only out of politeness but because her food had absolutely no resemblance to anything I ate before or after. Perhaps this is the kind of food served in Martian restaurants.

But let's get back to Earth. What did it look like in its early youth, when it still harbored no life at all?

The first seeds of life

Or perhaps it would be better to ask what happened to our ideas of early Earth after Pasteur made the theory of spontaneous generation obsolete and Darwin pointed out the existence of a common ancestor to all living creatures on the planet. Let us put panspermia aside for a moment (we discussed it at length in the previous chapter), for contrary to this notion, scientists of the second half of the nineteenth century began to consider the possibility that in some way life did originate right here, on our Earth.

As Greek philosophers used to speculate of the *arche*, the simplest matter out of which everything was created, so Weismann in 1902 came to an idea that there should exist something called the biophores, simplest and tiniest "carriers of life." In a way, this is a somewhat expanded version of Haeckel's theory of archegony (from Greek *arche*, ancient, and *genei*, to be born), first presented in 1866. According to the German naturalist (notable for his fundamental law of embryology stating that "ontogeny recapitulates phylogeny"), primordial forms of life were "homogeneous, structureless, amorphous lumps of protein."

But what does this mean? If we say that some form of matter is "homogeneous, structureless, amorphous" we are actually quoting a straight definition of chaos. This is obviously not the "deterministic chaos" of modern physics, or chaos of everyday conversation usage, but proper Greek Chaos, matter in its primordial form before it transforms itself into Cosmos, the structured world – as Ovid wrote at the very beginning of *Metamorphoses* (I.5–9):

> *Ante mare et terras et quod tegit omnia caelum*
> *unus erat toto naturae vultus in orbe,*
> *quem dixere chaos: rudis indigestaque moles*
> *nec quicquam nisi pondus iners congestaque eodem*
> *non bene iunctarum discordia semina rerum.*

> (Before the ocean and the earth appeared—
> before the skies had overspread them all—
> the face of Nature in a vast expanse
> was naught but Chaos uniformly waste.
> It was a rude and undeveloped mass,
> that nothing made except a ponderous weight;
> and all discordant elements confused,
> were there congested in a shapeless heap.)

The only difference is that Greeks, as well as Romans, had no idea what the constituents of their Chaos were, but Haeckel did. Unfortunately, in those times "protein" meant either a substance essential for life or nothing at all.

To do wrong

But how would this "protein" organize in the first living things, biophores or whatever? Even today there is a popular belief that this primordial organization could have appeared by pure chance, by a lucky intermingling of random chemicals. In my twenties I came into possession of a book *Sekularizacija medicine* (Secularization of Medicine) which was sent by its author to each and every Croatian scientific institution, editorial board of daily papers and magazines as well as to every embassy in a then much larger country, Yugoslavia (Croatia used to be one of the six federal units of former Yugoslavia).

I didn't know then, and I still don't know now, who was the author of the book, for he used the pseudonym Brljan, a nickname given to him by

3 900

hydrothermal vents
dominated surface

his mother (derived from Croatian word *brljati* – to be sloppy, do wrong). In a nutshell, he claimed to have created a living cell by mixing tap water with "elixir E," his label for a mixture of organic solvents used in carpentry. However crazy this might seem, in a strange way Brljan's experiments recapitulated similar experiments performed by serious scientists in the nineteenth century just as, according to Haeckel's "ontogeny recapitulates phylogeny," development of an individual replicates development of its species.

Such lines of reasoning, Brljan's as well as Haeckel's, led to a natural conclusion that life had something in common with crystallization. Life emerged from primordial chaos as an organized form of matter, just as crystals emerge from a solution, a chaotic mixture of solute and solvent molecules.

And there is yet another strong analogy. If every living creature has to feed in order to grow and replicate, then crystals have to be considered living things as well for they are "feeding" on their solute, "growing" by depositing solute molecules on their surfaces, and are even able to replicate, though not by their own volition. If we take a crystal of sugar and cut it into small pieces, each piece could become the seed of a new crystal if placed in a saturated sugar solution. This is a well-known chemical experiment, which everyody encountered in high school, with sugar, table salt, copper sulfate or alum. Chemists, of course, do the same, but their intention is not so much to grow beautiful crystals but to obtain their substances in a pure state, or to solve their structures by X-ray analysis.

Just as salt crystals are created through evaporation of salty ocean water, so the first living things would have appeared from some lucky mixture of organic as well as inorganic ingredients – by pure chance or through a miracle of God. Naturally, this led to first attempts at reproducing these primordial conditions or, at the very least, trying to make something resembling life in a test tube. It seems that the first scientist able to claim achieving this was the German chemist Traube in 1864 who placed a small crystal of copper sulfate into an aqueous solution of potassium ferrocyanide, i.e. potassium hexacyanoferrate(II). The precipitate in the form of semipermeable, osmotic membrane formed immediately, and the copper sulfate solution confined inside the membrane began to attract water through it. A droplet, or more accurately a bubble (vesicule), of the confined liquid began to grow and after a while the copper sulfate solution broke through the membrane and started forming another bubble – this process reflecting the growth and reproduction (doubling) of a living cell.

And those artificial living cells could even move; on their own! This development is not to be credited to Traube, but to his compatriot Bütschli who, in 1892, published a book *Untersuchungen über microscopische Schäume und das Protoplasma* (Research on the Microscopical Foams and Protoplasm). His "protoplasm" was nothing more than a drop of olive oil enveloped by a potash (i.e. potassium carbonate) solution, serving as a kind of cell membrane. These droplets even began to send out pseudopodia, as amoeba does. And there's more! In contrast to Traube, who created "inorganic" life, Bütschli's living cell was actually an "organic" one since olive oil is a mixture of organic compounds, mostly triglycerides of fatty acids and fatty acids in their free, unbound state. Free fatty acids? These acids were obviously neutralized by potassium carbonate to form soap, which then engulfed the droplets. Perhaps you have unwittingly accomplished exactly the same while washing dishes.

The most ardent disciple of this chemistry (or alchemy?) was the French scientist Leduc at the beginning of twentieth century. He placed a piece of fused calcium chloride into a saturated solution of potash and potassium phosphate. Again, as with Traube's experiments, an osmotic bag was formed and, what is more, all kinds of "live forms" appeared resembling mushrooms and algae, not to mention other members of the Plant Kingdom. And those forms were even able to grow!

Such games with osmotic forces are popular even today, but not as a method for creating life in a test tube. If you take fused calcium chloride, as well as crystals of copper(II) and iron(II) sulfate, iron(III) chloride, or just any soluble heavy metal salt which may be found on a shelf in any chemical laboratory, and put them into a diluted solution of sodium silicate ("water glass"), you would obtain similar structures closely resembling grasses and corals. In this case, the osmotic membrane is formed by precipitating silicates, and so "plants" found in these "chemical gardens" are made of stuff closely resembling concrete. (It is possible to grow a "chemical garden" in a test tube, as well as in a vessel of any size. In 2000 I prepared one such garden, for an exhibition of popular science, by mixing ten liters of water glass solution with twenty liters of deionized water, and "seeding" it with crystals of ten different salts.)

In that same year, 1907, when Leduc published the results of his experiments on creation of artificial life, Kuckuck published a book with a highly presumptuous title *Lösung des Problems der Urzeugung* (A Solution for the Problem of Autogeneration). And how did he solve this problem? He simply exposed a mixture of gelatin (that is protein), glycerol and table salt to radiation released by radium. He claimed that after 24 hours a peculiar culture appeared on the gelatin, quite similar to bacterial colonies growing on agar in Petri dishes. These "living cells" were able to grow and divide, and even appeared to exhibit other manifestations of life. Now that would be too much even for Brljan ("Wrong-doer")!

Such experiments may now appear quite naive, of course, but it all depends on one's viewpoint, on the scientific paradigm dominant at the time. If life can be created by simply mixing chemicals, as Darwin's "warm little pond" suggests, then every lifelike structure obtained by some kind of mixing is a milestone, however small, leading towards the final solution of the greatest mystery of natural history: the emergence of life on our planet. Nevertheless, life obviously didn't emerge from copper sulfate and potassium ferrocyanide or calcium chloride and potassium phosphate, but from chemicals which are found in today's living creatures – sugars, alcohols, fatty and amino acids, proteins. And how were these substances first formed? Chemistry of the nineteenth century had its answer to this question as well.

Magmatic chemistry

Dugi Rat, a small town on the Adriatic coast a few kilometers west of Omiš and yet another small town where Marija perfected her *pašticada*, used to house the pride of Yugoslav chemical industry: a huge calcium carbide chemical factory. Since my father had relatives in Dugi Rat who were employed by the factory, he took an opportunity to show me the process of producing the well-known *karabit* (as we called calcium carbide). I learned that it was made in electric-arc furnaces by smelting quick lime (calcium oxide) with coke. I also learned that a new stone quarry had just been opened solely to furnish the factory with limestone, which was then converted into lime in factory furnaces. Even the new hydroelectric power plant at a nearby river was built especially to provide the factory with electricity, long before this symbol of national prosperity reached rural homes in the factory's neighborhood.

<table>
<tr><td>3 810</td><td>3 800</td></tr>
</table>

the oldest rock with possible organic
origin (western Greenland)

origin of life (?)

But there were other sights to be seen in this small coastal town. It was a dirty-looking place choking on grey dust and noxious fumes from the gas released by carbide's reaction with water – the chemical reaction which was the *raison d'etre* for the whole enormous complex. When calcium carbide reacts with water, acetylene (ethyne) is produced; a hydrocarbon gas with a simple formula C_2H_2. This gas was used not only for welding and making funny noises during New Year's Eve celebrations and similar occasions, but also for producing plastics, as well as many other chemicals and chemical products, including synthetic rubber (buna). Nippon Chisso, a Japanese calcium-carbide factory situated in Minamata bay used to produce numerous chemicals whose synthetic production routes all began with hydration of acetylene on mercury catalyst. The tragic Minamata story is well known: after 1,435 estimated deaths from mercury poisoning, the factory was closed down, just like the one in Dugi Rat. They were both built on the coast to facilitate coal transportation – however, as they were being built no one gave any thought to the beauty and the wealth of the sea, not to mention health of the people living nearby.

This reaction of calcium carbide with water introduced me to the chemistry of carbides in general. Calcium carbide reacts strongly with water at room temperature producing only one hydrocarbon, ethyne. Other carbides are much more chemically inert, but they may also be converted into hydrocarbons through reaction with water, now in the form of steam. And it is exactly this process which paved the way for a peculiar theory of the origin of organic matter on our planet, and even the origin of life itself.

It is an old idea. The first chemists initially observed that various mixtures of hydrocarbons would form as a result of dissolving cast iron in hydrochloric acid. Of course, as we know from elementary chemistry, hydrogen gas is released by dissolving iron, as well as other metals, in an acid. However, as well as the metal itself, cast iron contains iron carbides. The same holds true for ferromanganese, an alloy of iron and manganese. It too contains a few percent of carbon. Furthermore, early chemists obtained liquids resembling petroleum by treating cast iron or ferromanganese with hydrochloric acid or superheated steam. After repeating and corroborating their experiments, in 1877 Mendeleev put forward his theory of mineral origin of petroleum: "As the igneous rocks were folded, cracks must have been formed which at the crests opened outwards while at the depression they opened inwards. Both these types of cracks became in time filled in, but the more recent the origin of the mountain the more open these cracks be, and water must have entered through them into the Earth's interior to such depths as would be impossible normally from a plane surface."

But we don't even have to agree with this supposition in order to accept the theory of mineral origin of petroleum. There is quite enough water in Earth's mantle, the molten rocks on which continental plates float, to account for it. This water rose to the surface through volcanic eruptions, and it was evidently hot enough to convert carbides into hydrocarbons.

This is very simple chemistry indeed, and it could presumably lead to discovery of enormous amounts of oil somewhere deep within Earth's crust. Mendeleev's theory achieved wider popularity somewhat later, in the twentieth century, helped by Thomas Gold, an American astronomer who independently came to the idea of "abiogenic formation of fossil fuels." His notions fell on deaf ears of geologists in general though. Oil deposits have formed during the Mesozoic era, they claimed, from organic matter in shale sediments, thus ruling out the possibility of any other origin or any other kind of petroleum deposits – despite occasional findings of hydrocarbons in rocks very much older than Mesozoic.

But let us put aside, for a moment, such ideas that oil deposits are being continuously formed and that oil is, in effect, a source of renewable energy (for carbon finds its way into Earth's interior by carbonate deposits on continental plates) and let us examine our home planet at its very beginning. Earth was formed, as all other planets were, by accretion of smaller celestial bodies, planetesimals. Gravitational energy is released by this process as heat, and consequently Earth used to be a much hotter place. So much so that it probably spent its first few hundreds of millions of years in a molten state. The first igneous rocks began to form when rivers of lava ran on the surface of Earth and carbon, chemically bound or dissolved in molten rocks, reacted with water vapor. Hydrocarbons inevitably formed, as the first organic matter on the planet.

Hunting the live molecules

"He who is tired of hydrocarbons, is tired of chemistry," pronounced my elderly colleague to a much younger chemist when asked why after thirty years of dealing with hydrocarbons he still hadn't made any attempt at applying his theoretical methods to other chemical compounds, the implication being that they would not work well with more complex systems. Hydrocarbons are the simplest organic compounds indeed, and all other carbon compounds may be considered their derivatives. This means that it is possible to consider any organic molecule as a modification of some hydrocarbon molecule or other or, better yet, as a hydrocarbon following substitution of one or more of its hydrogen atoms by appropriate functional group or groups. This is reflected in chemical nomenclature, where all kinds of suffixes, prefixes, numerals and Greek letters indicate the type and mode of binding to a hydrocarbon skeleton. A lengthy and outlandish chemical name, as lengthy and weird as it may be, is easily translated by a chemist into something like: take this or that hydrocarbon and substitute hydrogen atoms on this and that carbon atom with this or that group of atoms.

It is easy to write formulae on a sheet of paper, but it stops being easy when they have to be formed into actual substances, especially if synthesis has to start from the very beginning, the parent hydrocarbon. Namely, hydrocarbons are very inert substances. This is especially true for saturated hydrocarbons or alkanes. On the other hand, unsaturated hydrocarbons, especially those with triple bonds, are much more prone to modifications. We already mentioned the simplest hydrocarbon in the series, ethyne – C_2H_2, the gas released by reaction of calcium carbide with water. This reaction was essential for chemical technology of the twentieth century as it allowed the use of carbon derived from coal, via its primary conversion to calcium carbide, for production of virtually all organic chemicals. The second key reaction in this obsolete technology is reaction with water, which is in this case the reaction of water and ethyne. As a result of adding water to its triple bond, an aldehyde is formed, acetaldehyde ("aldehyde of acetic acid") or ethanal. Its formula is quite simple: CH_3CHO.

On the other hand, acetaldehyde cannot be obtained just by simply dissolving ethyne in water; a catalyst is needed. One substance commonly used for this purpose is mercury sulfate, a very toxic compound indeed and the culprit behind the disaster which struck Minamata bay. As mercury accumulated in the food chain its levels in humans, mostly fishermen and their families who prepared their meals from fish caught in the bay, rose up to a thousand times above their normal values. Such a massive case of poisoning led to an official number of 1,435 deaths out of 2,265 victims.

It is impossible to imagine ethanal forming on early Earth through this process; there was no mercury sulfate in the primordial ocean and even less quick lime, coal and coke, not to mention a conspicuous absence of a working electric furnace. Nevertheless, as Oparin suggested in his book *The Origin of Life*, first published in 1936, ethyne might have still formed. If methane is heated to about 1,000 ºC it is converted to ethyne without the need for any catalyst. To convert it further into ethanal a catalyst is still necessary, but it doesn't need to be mercury sulfate. Iron oxides can do the job as well, but they require rather high temperatures, approximately 300 ºC, as Russian chemist Chichibabin demonstrated in 1915. All you need are hot rocks and plenty of superheated steam.

"Considerable quantities of various oxidation products of hydrocarbons, such as alcohols, aldehydes, ketones, and organic acids must have originated as a result of such transformations on the Earth's surface," wrote Oparin. "In the above reaction described, as Tchitchibabin [Chichibabin] points out, if the heated moist acetylene gas contains ammonia, it is possible to observe with the naked eye the formation of a crystalline precipitate of an aldehyde-ammonia; i.e., under these conditions ammonia very rapidly combines with acetaldehyde formed by hydration. Similarly, other oxidized derivatives of hydrocarbons (the above mentioned alcohols, aldehydes, and acids) can enter into a variety of reactions with ammonia giving rise to ammonium salts, amides, amines, etc." In other words, what humans in Oparin's times were producing from ethyne ("acetylene gas"), Nature had produced herself from the same substance in some remote age.

But what happened next? A late colleague of mine, who used to teach chemistry in one of Zagreb's high schools, liked to talk about his days as a student of chemistry immediately following the end of World War II. The lecture-room was located in a stable and students used to sit on rough wooden blocks. Lectures commenced at 7 A.M. sharp, and their professor, a high Communist official and former partisan fighter, arrived to deliver his lectures in a luxury car driven by his personal driver. And what about his exam questions? There was only one: "Tell me of the reactions of *adelhydes*." Aldehydes, adelhydes, adelyde, Adelaide ... Though most professional chemists aren't even aware that aldehyde is actually an abbreviation for

3 690
|

---

fossils of Fe^{2+}–oxidizing
organisms (Quebec, Canada)

alcohol dehydrogenatus (dehydrogenated, i.e. oxidized alcohol), all of them know perfectly well that aldehydes are the key to organic synthesis.

Reduction of aldehydes leads to alcohols while their oxidation produces acids. It is possible to do both in just one step, simultaneously synthesizing alcohol and acid of the same aldehyde by simply adding a sodium hydroxide solution. It is the well-known Cannizzaro reaction, named after Italian chemist Stanislao Cannizzaro, a revolutionary as was my late colleague's esteemed professor, but this one did his revolutionizing in the turbulent year of 1848. Aldehyde molecules may also join together, thus forming polymers with many hydroxyl groups which are known as polyvalent alcohols.

This reaction, known as aldol condensation, is catalyzed by a different set of inorganic salts such as zinc chloride, as Wurtz discovered in the nineteenth century, and by milk of lime, as found by Russian chemist Butlerov in 1861. If you dissolve the simplest aldehyde methanal, or formaldehyde ($HCHO$) in water and then add milk of lime (suspended calcium hydroxide), sometime later and with a bit of luck you may notice formation of crystals of simple sugars. This is the famous formose reaction, which is not named after Formosa (an old name for Taiwan) but is actually an abbreviation for "formaldehyde → glucose," or any other "ose" (allose, altrose, arabinose, xylose... being the names of simple sugars). Modern analysis revealed the compositions of the products of formose reactions as 10% C_4, 30% C_5, 55% C_6, 5% > C_6, where C_4, C_5, etc. refer to the number of carbon atoms in sugar molecules. If boric acid in the form of its calcium salt (also known as mineral colemanite) is added to the solution, ribose – the key constituent of RNA – is formed with a high yield.

Similarly to how hydrocarbons are formed, i.e. by reaction of carbides with superheated steam, ammonia (NH_3) may be obtained, namely by reaction of nitrides with water. By means of reaction of ammonia with aldehydes, as well as with the products of formose reaction, certain rather complex compounds may form, some even quite similar to nucleobases.

This branch of chemistry, which formed in the middle of the nineteenth century, is still very much alive today; many chemists still study the mechanisms and products of apparently simple systems of formaldehyde in alkaline solutions, with or without ammonia. Those are the chemists who are trying to reproduce conditions on the early Earth and to uncover influences of various factors on formose reaction. "Summarizing briefly what has been said, we can draw the following conclusions: Hydrocarbon derivatives, such as alcohols, aldehydes, organic acids, amines, amides, etc., undergo important transformations when their aqueous solutions are allowed to stand," wrote Oparin in his famous book. "As a result, numerous high molecular compounds, similar to those present in living cells, may appear in aqueous solutions of hydrocarbon derivatives on long standing."

Long standing – but how long? It is not unimaginable for an aqueous solution to stand a hundred, a thousand, even a million years, because we are dealing with geological time scales here. The real problem is how those "numerous high molecular compounds" got themselves arranged into the first living cell. By chance? No chance, no chance at all!

God, life, Communism

Communism is a dirty word nowadays. And I don't have the slightest intention of defending it, especially since my family did their fair share of suffering in this distorted political system. However, if we see Communist ideology only as a basis for terror and political repression, we may be taking too narrow a view. To equate Marxism with Lenin, Stalin, Ho Chi Min, Fidel Castro or Che Guevara is the same as identifying Islam with Osama Bin Laden and ISIL, Christianity with Cromwell, Spanish Inquisition and witch-hunting, Enlightenment with Robespierre's dictatorship and Napoleon's war crimes, and – why not? – Wagner's music with concentration camps. But what is this "Communism" thing, or to be more precise, what is Marxism?

In its very essence Marxism is a radically materialistic interpretation (or reinterpretation) of Hegelian philosophy, known as historical or dialectical materialism. Everything on Earth is to be regarded from the perspective of time and change – nothing stays the same forever. From another viewpoint, there are no miracles in nature; new phenomena emerge only as results of a new, higher complexity. Quantity results in (new) quality, says one of the basic laws of dialectics; the "soul" is only a consequence of a multitude of neurons working together in a small volume – that of the human brain. (I hope mine works well.)

"Such theories, which treated the origin of life in mechanistic fashion, were bound to offend the religious, but Oparin lived in a nation that, after

1917, was officially atheistic," wrote Asimov in his *Biographic Encyclopedia of Science and Technology*, adding: "He had nothing to fear from governmental piety." Moreover, it is hardly imaginable for someone to come to the same conclusions as the Russian scientist, or to accept his approach to solving the problem of life's origin, without dialectical materialism, the doctrine originally introduced in the book *Dialectics of Nature*, written by Friedrich Engels, the first supporter and a close friend of Karl Marx.

Dialectical materialism is opposed to idealism (theism, vitalism, etc.) as well as to crude "mechanistic" materialism, so it is a bit misleading that Oparin "treated the origin of life in mechanistic fashion," as Asimov claimed. "Life is a form of the existence of protein bodies," said Engels, but this does not mean that "life is a protein body" and nothing else, as a "vulgar" (i.e. mechanistic) materialist would brashly claim. Through a particular organization of simple molecules new properties will emerge – dialectical materialism is essentially a philosophy of complex systems.

And that's the way it is with the origin of life. Life was not ordained by a miracle of God or by action of some mysterious vital force (as "idealists" proposed) and neither did it emerge by sudden appearance of a "live molecule" or "live molecules" (as "vulgar" materialists believe); it came to be, just as Oparin said, by stepwise evolution: "All these difficulties, however, disappear if we discard once and for all the above mechanistic conception and take the standpoint that the simplest living organisms originated gradually by a long evolutionary process of organic substance and that they represent merely definite mileposts along the general historic road of evolution of matter."* The proper name of his book, as well as that of the topic it deals with – as suggested by Sergius Morgulis, its English translator – should be "life's coming into being," rather than "the origin of life."

Aleksandr Ivanovich was obviously not just some dissident, but an ardent believer in Communism. In the year of the October Revolution, 1917, he graduated biology (plant physiology) from Moscow State University and began his scientific career to eventually become a professor of biochemistry. His main field of interest was enzymology, the science of "living

---

*Or more explicitly, in the second (1941) Russian edition of his book (p. 48) "(dialectical materialism) states that life came into being as a new quality in the process of historic development of matter."

proteins" – and this quite naturally led him to his theory of the origin of life. "In general, the students considered A. I. Oparin an Olympian, almost a celestial being," remembered M. S. Kritsky, his former student. "But his arresting appearance (with all the attributes characteristic of an Academician according to cinematographic and literary standards) and the main field of his scientific interests, the problem of the origin of life, upheld this attitude of ours."

Communists were obsessed with heroes, outstanding and superior personalities (according to their standards). During the Communist era there was only one organic chemist in Croatia, the above mentioned professor who was in the habit of arriving to his stable-lectures in a manner befitting a head of state. There was only one writer, only one painter, only one bio-medical scientist, etc; other writers, painters, and scientists were virtually non-existent. There were "popular masses" and their leaders who led them through the "intermediate" state of Socialism to the Ultimate Ideal Communism, a political system in which everyone would work and spend as much money as they pleased, if the concept of money as such existed at all anymore. This was the political constellation in which Oparin put forward his theory on the origin of life in 1922.

This theory launched him into the orbit of Communist Olympus. In 1935 Oparin founded the Biochemistry Institute by the USSR Academy of Sciences; in 1939 he became a Corresponding Member of the Academy, and in 1969 a Hero of Socialist Labor. In 1957 Oparin organized an International Symposium on the Origin of Life in Moscow, and went on to organize and notably attend many similar symposia afterwards. He was a very popular figure not only in Russia, but also abroad, in the "capitalist" world. He made acquaintance with Salvador Dali, a notable Spanish painter who also held an intense interest in science. Even the President of Pontifical (Vatican) Academy of Sciences honored him with a commendation: "You know, Professor Oparin, I really admire how you brilliantly managed to reveal divine Providence..."

Was life on Earth really initiated by divine Providence or did the laws of dialectical materialism play a crucial role? It depends on your viewpoint. Something happened in the "warm little pond," that is certain. But what? This is a scientific problem rather than a philosophical or theological one (not to mention political).

Droplets of life

"It was believed previously that organic chemical substances were created and changed in the living beings by action of a peculiar mystic vital force, but nowadays we come to an evident conclusion, drawn from a very few properties observed in colloids, that the colloid state of chemical substance of an organism, first of all that of the proteins in a living cell, makes possible... to put in effect such processes we are not able to reproduce in our laboratories and factories," wrote Croatian biochemist Fran Bubanović in 1918, four years before Oparin first spoke about his hypothesis on the origin of life at the meeting of the Russian Botanical Society. This sentence echoes a common opinion among biochemists of his time: in its very essence life is a colloid phenomenon, or to be more specific, all processes in a living cell should be attributed to changes of the state of "live" protein colloids. But what exactly are colloids?

The term "colloids" came from their discoverer, a nineteenth century Scottish chemist Thomas Graham. While studying diffusion he observed that its rate is not the same for all substances. In 1831 he noticed that the rate of diffusion of a gas is inversely proportional to the square root of its molecular mass – a law which can be found in any textbook on physical chemistry (Graham's law). But what about aqueous solutions? If you slowly and carefully pour demineralized water over a copper sulfate solution contained in glass cylinder, you will obtain two separate liquids: a blue one at the bottom and a colorless one at the top. After a few days at rest they will take on the same blue color due to diffusion of copper ions. (The same could be demonstrated with red wine by pouring, very slowly, this wine on top of water – and you may have a nice glass of *bevanda* after the experiment.)

In one such experiment Graham placed a sheet of parchment between water and the solution. He noticed that some solutes passed through the parchment while others did not. And those which did not pass – like glue, gelatin and gum arabic – diffused very slowly. In 1861 this led him to the conclusion that there were two kinds of solutions: colloids (those similar to *kolla*, Greek for glue, i.e. gelatin) and crystalloids, those which can be crystallized. We now know that colloids' particles are much bigger than those comprising crystalloids. Simple metal ions have diameters in the range of a few tenths of a nanometer (1 nm = 10^{-6} mm). Colloid particles, on the contrary, are up to a tenth of a micrometer (10^{-4} mm), or a hundred nanometers, across and their lower size range was put conventionally at one nanometer.

These particles come in two kinds. Many of them are big molecules, macromolecules, i.e. polymer molecules (note that hemoglobin molecule is five nanometers across), others are minuscule lumps of insoluble, usually inorganic substances (silver bromide, sulfur, silver, gold, etc), better known as nanoparticles ("particles of nanometer size"). But both kinds of particles are stabilized in a solution by two forces – electrostatic force and hydration force.

It would be logical to assume that colloid particles aggregate naturally in order to minimize their surface area exposed to water, but in reality this is prevented by their repulsion of one another. This happens because all particles in the solution have the same charge, whether positive or negative. Moreover, this repulsion is further augmented by a cushion (or bumper) of tightly bonded water molecules (hydrophilic colloids) surrounding the particles. In protein colloids both forces are in operation, and that is why tofu (soy bean cheese) can be prepared from soy milk (which is actually a colloid solution of soy proteins) in two ways: by mixing it with vinegar (to neutralize the charges of protein molecules) or by adding magnesium salts (to destroy their "water cushion").

If life is a "colloid state of chemical substance of an organism," as Bubanović claimed, or more explicitly "protoplasm is a highly complex colloidal system" (S. Morgulis), then the principal question on the origin of life is how did those "highly complex colloidal systems" emerge from primordial ocean bouillon, namely Haldane's "hot thin soup?" It is well known that hydrophilic colloids exist in two forms, as a dissolved liquid (sol) or in a solid (gel) state. Those states turn from one into another by heating or cooling, shaking, standing still, etc. And those processes are usually not accompanied by a chemical transformation; solid yogurt, popular in America, is nothing but a gel state of its liquid variant (sol) which is more commonly drunk in Europe. And yet there is a third, less known, state of colloid systems on which Oparin founded his theory. It is coacervate, from *coacervatio*, Latin for accumulating or collecting (or *coacervatus* – accumulated, collected).

3 510   3 500

the split between bacteria
and archaea occurred

Coacervates are complexes of colloid particles, which are formed when forces of attraction and repulsion among colloid particles arrive at a peculiar state of equilibrium resulting in their accumulation in a certain region of the colloid solution. It then separates into two layers, fluid sediment, rich in colloid particles, and a liquid layer free of them. This "fluid sediment," the concentrated ("accumulated") colloid solution, may appear in the form of minute droplets floating in clear liquid.

A droplet of coacervate strongly resembles a living cell, which is also "a highly complex colloidal system." If this were the only fruit of Oparin's mind, he would be judged not much better than Leduc, who succeeded in producing "artificial" mushrooms and algae, not to mention Brljan's experiments with his "elixir E" (here *E* is not an emphasis of the word elixir, but it was placed there in honor of Brljan's wife Elisabeth). Coacervation of colloids enabled separation and, importantly, individualization of colloid solution from its surroundings; or in Oparin's words: "The formation of coacervates was a most important event in the evolution of the primary organic substance and in the process of autogeneration of life." And even: "Before this event organic matter was indissolubly fused with its medium, being diffused throughout the mass of the solvent. But with the formation of coacervates organic matter became concentrated at different points of the aqueous medium and, at the same time, sharp division occurred between the medium and the coacervate."

This "sharp division" did not only enable a separate development of every coacervate droplet which inevitably led to natural selection and evolution, but it also made possible the transfer of organic, as well as inorganic substances throughout the boundary between the droplet and its surrounding medium. And so it happened that the first living cell was formed, with its colloid protoplasm and a cellular membrane.

But have we misunderstood Oparin? Coacervate droplets which formed in the primordial ocean were not alive because there was no, and nor could there be, an indisputable boundary between the living and the non-living. The living cell did not emerge all at once but "by a long evolutionary process of organic substance," as he repeatedly claimed. The formation of coacervate droplets was just one step, however important it may be, in the evolution of our biosphere.

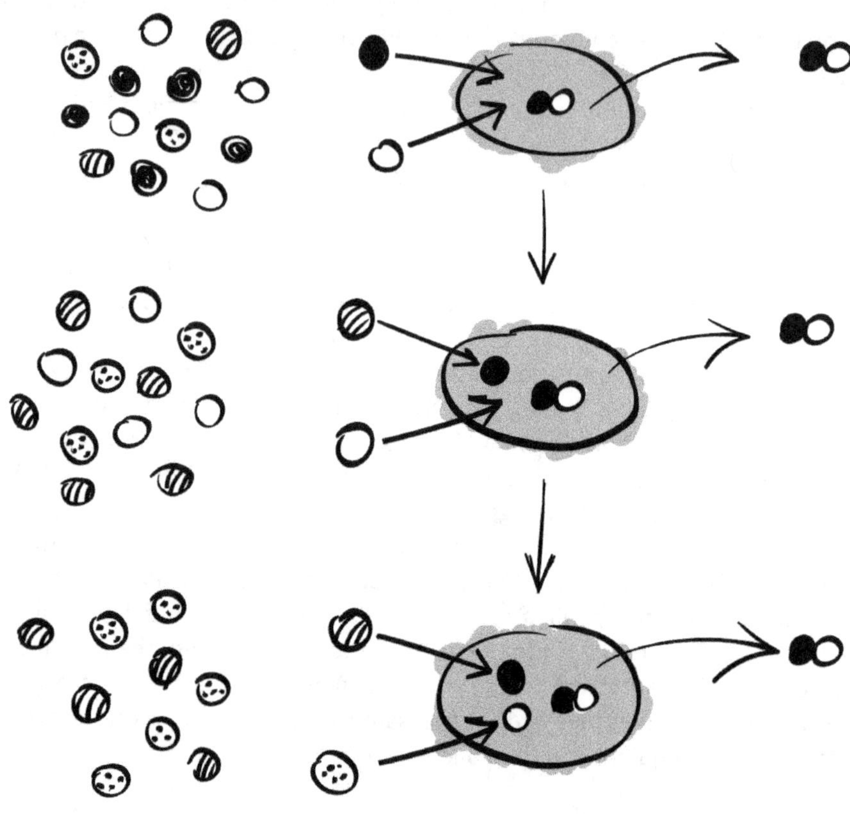

Evolution of protocells, i.e. coacervate droplets, after Oparin: the first catalytic reaction (○ + ● → ○●) turned simple droplets into the more complex ones as ○ and ● gradually depleted from the environment. And so evolved the first metabolisms.

Evolution of colloids

Droplets were formed, reformed, destroyed and then formed again... Their constitution did not vary much if they were formed in the same and *of* the same *bouillon*. Even after they were formed, the concentration of salts and organic matter within them didn't differ from that on the outside and so it went on – forever. How could something resembling life emerge from that?

3 450

fossils of sulfur–reducing
bacteria (South Africa)

But it is not all so hopelessly simple. First of all, coacervate droplets tend to absorb various substances from their surroundings at various rates. By this process the composition of a droplet begins to respond to changes in its environment. Even greater changes are to be expected if some kind of chemistry is present. If there was some primordial protein or similar molecules in the droplet capable of consuming some substance from their neighborhood by converting it into something new, the chemical composition of the droplet would change correspondingly. If substance A was converted into substance B, then concentration of substance A would drop and concentration of substance B would rise. As diffusion through the boundary of a coacervate droplet tends to equilibrate concentrations inside and outside it, substance A will diffuse into and substance B will diffuse out of the droplet.

But then again, what would happen if that substance B is composed of colloid-sized molecules? In that case it would get accumulated inside the droplet. As a consequence, the droplet would increase in size to eventually divide itself into two virtually identical "individuals." And so droplets with "good" chemistries would have better likelihood of surviving and reproducing – in their way, of course – than those with "bad" chemistries, or no chemistry at all. "In this way a natural selection of coacervates originated in its most primitive and simplest form, only the dynamically most stable colloidal systems securing for themselves the possibility of continued existence and evolution." This is the essence of Oparin's theory. But is there any experimental proof?

Oparin prepared many coacervates, such as those obtained by mixing a colloid solution of gelatin with colloids of gum arabic, egg lecithin or protamine. He found that many organics were almost completely removed from the surrounding media by action of his coacervates.

Nevertheless, his attempts to reproduce a catalytic, i.e. enzymatic reaction inside the droplets were not very successful because, among other things, in those times nobody knew how to prepare enzymes in their pure state. Although it would be quite easy to create coacervate droplets doped with purified enzymes today, chemists have virtually lost all interest in such experiments. Why?

Protoplasm is not merely just another colloid solution, as Oparin's fellow scientists thought; "a complex colloid system, coacervate intermicellar substance, in which are present fibrous colloide micelles..." We now know that enzymes are usually bonded to cell membranes working together to produce something that, at the first glance, looks like a product of just one reaction, the work of a single enzyme (as in a test tube). The Russian biologist was conscious of the problem, but he had no means to solve it: "The artificial coacervates, which we can obtain by mixing together different colloidal solutions, and the analogous formations, which have originated in the hydrosphere of Earth, do not have and never had have the highly efficient physico-chemical organization described above (i.e. that of the living cell)." But the general idea persisted.

First of all, first living organisms were heterotrophic. This means that they fed on complex organic substances found in their environment in order to exploit their chemical energy. The first catalyst gave rise to the first metabolism (A→B), which later developed into C→A→B, D→C→A→B, etc, when "nourishments" A and C were gradually depleted from the droplets' environment.

But above all, Oparin did not provide any explanation of how new genetic machinery developed later on, nor did he explain how the acquired structure of coacervate droplet could persist during innumerable doublings. Nevertheless, the Russian scientist did initiate a new approach to the problem of the origin of life on our planet. Life originated from very simple systems, pioneer organisms, as they are called today, and not from something resembling modern bacteria or viruses. Biological evolution is an evolution of systems, and not an independent evolution of molecules, cells or organs. There has been no evolution of proteins or nucleic acids as such. They evolved and still evolve by adapting to their environment, as their environment evolves by adapting to them and to their molecular structure. This is the fruit of dialectical materialism, whatever you might think about Communism otherwise.

Did Oparin succeed in preparing his French soup of life after all? Hardly, but he made a magnificent *borscht*, no matter what I happen to think of this dish myself.

3 390

Mill. yrs from present

3 360

Mill. yrs from present

Did it all begin in the air?

3 330

Mill. yrs from present

4. MILLER'S BEER

In 1953 a young American scientist synthesized amino acids in simulated primitive Earth's atmosphere − and solved the enigma of the origin of life. But did he really do it? The problem is, no doubt, much more complicated than that.

"In the beginning, God created the heavens and the earth." But what *was* this "the beginning?" In the beginning "the earth was without form and void, and darkness was over the face of the deep. And the Spirit of God was hovering over the face of the waters." In the beginning there was a vast, vast ocean and the sky above it was devoid of any bird, any fly, any yeast, any germ, or any bacterium. The wind, "the Spirit of God" moved upon it, but this wind – as science now claims – was of a somewhat unusual composition.

It wasn't composed of nitrogen, oxygen, carbon dioxide, argon and traces of other noble gases, as is the air we breathe; in fact it was a rather strange mixture of methane, ammonia, hydrogen and water in the form of vapor. "And God said, 'Let there be light,' and there was light." Yes, it seems we have overlooked something – light. And yes, there was plenty of light, for there was no ozone layer to protect the Earth from Sun's ultraviolet radiation. This radiation certainly did its job. "And God saw that light was good" – from the light and "the Spirit of God" the first life emerged.

This report from the Holy Scripture leads us to the third recipe: "Take a mixture of methane, ammonia, hydrogen and its oxide, also known as water – the infamous 'dihydrogen oxide' – expose it to UV radiation and you will get..." Hmm, not life really but rather a mixture of amino acids and many other very important chemicals with which you may prepare Oparin's bouillon or something quite similar. This was the young American chemist Stanley Lloyd Miller's recipe, which he used to prepare his first serving in 1953. It was the year in which his countryman James D. Watson, together with his British colleague Francis Crick, announced disclosure of the secret of life by solving the structure of DNA, the molecule of the gene.

3 300

moderate climate
on Earth

Miller's soup or Miller's beer? Beer perhaps, because – as one commercial claimed – beer is prepared from the four elements, at least according to ancient Greeks. They believed that everything was made, or more accurately, everything had its origin, in water, earth, air, and fire, just as genuine beer is created from barley, hop, yeast, and water. Miller made his brew from four ingredients as well: methane, ammonia, hydrogen, and water.

German beer – American beer

Beer can also be made, as you probably know, from wheat; and if you bother to read the minuscule letters on the bottle, you may discover that it can contain maize as well. But according to German law, beer may only be brewed from the four great components – barley, hop, yeast, and water – and nothing else. Additionally, a German will usually order just "beer," rather than a brand name such as "Budweiser," or "Amstel," or "Löwenbräu." As Austria is a member of Germanic cultures and for nearly a thousand years Croats were subjects to Austrian kings, in beer halls, bars and restaurants we also order simply "beer." The waiter may ask you for your order. "Beer." "Which beer?" "Which beer do you have?" And then he would count off all kinds of beer he could remember. After a while you'd say: "The third one – what was its name?" "Velebitsko," he answers. "That's it!" you'd exclaim. "I can't believe you have it! This is the best beer brewed in Croatia." And it really is. It is brewed in a mountain region, with spring water running from Mount Velebit which gives it a very refreshing taste. Because of its limited production, I couldn't find it even in Gospić, a town just a few miles distant from the place where it is produced.

There is a lesson to be learned here. During World War II a German spy on a covert mission in the US ordered a beer by simply calling for "beer" and was immediately arrested as a suspicious alien. With this anecdote firmly in my mind, during my stay in the US as a Fullbright scholar back in the final days of the Cold War, I used to order my favorite in the States by calling for "Budweiser." Nonetheless, I still refused to pronounce the name as Americans do, "budveezer," but in the proper German way, even going so far as to use the proper German accent (their *r* differs from both English and Croatian *r*). People probably thought I was German, but thankfully those times were long gone and I managed to avoid being shot.

"How do you like it in America?" was a question I was asked many times by Americans when I visited there to do my postdoc on one very interesting molecule. "Everything is the same – but different!" used to be my standard answer. And it really is so. Americans are indistinguishable from Europeans, but Europeans they are not. American houses look like houses most Europeans live in, but American houses are generally constructed from wood rather than bricks, as is the custom in most of Europe. American front doors usually open to the outside rather than to inside as in Europe. American English is subtly different from British English and the similarity only serves to emphasize the differences. (My son spoke with a "proper" RP accent taught in Croatian schools and one of his school friends declared that he "sounds like my uncle from London.") And finally, we come to American beer – which is not beer at all, at least not according to European standards...

Before commuting to my job I liked to have a drink from a vending machine found on the way to my institute. One day my attention turned to a colorful beverage can I didn't notice before. What is this? Beer? Yes, really – root beer. But it was not beer at all. To my European palate it tasted of tooth paste, and I found it utterly disgusting. How on the earth can someone dare to call this thing "beer?!" But my little daughter held a very different view. She liked this root beer so much that when she grew older she asked her American friends to bring her a bottle of this "marvelous" drink from America. She kept this bottle in the refrigerator for a long time, and drank from it sip by sip. *De gustibus non est disputandum*. Tastes are not to be discussed.

On the other hand, the origin of life is a subject which should be discussed. Why did Stanley Miller put methane, ammonia, hydrogen and water in his apparatus, and nothing else besides? And why did he expose this mixture to electric sparks?

Sparks in the gas mixture

Stanley Lloyd Miller was, plainly speaking, a doctoral student of Harold Clayton Urey, a notable American chemist who was awarded the 1934 Nobel Prize in chemistry for the discovery of heavy hydrogen. This led many authors to refer to the Urey-Miller or Urey/Miller experiment, and even begin writing stuff like "they observed," "they found," "they concluded" and "Miller and Urey imitated the ancient atmosphere on Earth," and so on. In other words, the esteemed professor had a great idea and his obedient student helped him pursue it. But nothing even remotely like this actually happened!

The actual truth is that Stanley attended Urey's lectures at the University of Chicago and learned about the professor's hypothesis on composition of primordial Earth atmosphere. As he happened to be looking for a suitable subject for his dissertation, in September 1952 he asked Urey about a possibility of experimental proof for prebiotic synthesis from the mixture of gasses modelled according to his hypothesis. This was a rather tricky proposition for Urey, and at first he did not go along with his student's suggestion. In his opinion, which was consistent with the one held generally, a scientific problem suitable for PhD thesis has to be solvable in a clear and relatively easy manner. The student persisted however – and so, at the age of 23, he wrote a thesis that shook the world.

"The formation of the solar system had been discussed by Urey on the basis of a model that starts with a dust cloud containing the elements distributed in proportion to their cosmic abundances," runs the first sentence of Miller's paper published in the May 1955 issue of the *Journal of the American Chemical Society*. "Two of the principal conclusions of his work are that the planets formed at low temperatures (<300°) and that they had reducing atmospheres in their initial stages of formation. The reducing atmosphere would consist of methane, ammonia, water and hydrogen instead of the present oxidizing atmosphere of carbon dioxide, nitrogen, oxygen and water."

Reducing atmosphere, but why of that particular composition? Oparin too spoke in favor of such an atmosphere, Miller noted, but the notion of a primordial atmosphere on Earth devoid of free oxygen could be traced (as Oparin also said!) back to the nineteenth century scientist Svante Arrhenius, a Swedish chemist with whom we had already made acquaintance as one of the fathers of panspermia theory.

But Oparin had no clear idea which gases were present in Earth's atmosphere when it was first formed. It was composed, he said, of "super-heated aqueous vapor, some nitrogen and other heavier gases." Carbon and nitrogen compounds were hardly present, because these two elements had been combined, as described in the previous chapter, with metals. All organic stuff on Earth originated later, when carbides and nitrides began to react with super-heated water vapor to produce hydrocarbons and ammonia. But according to Urey, primordial Earth atmosphere was much colder than previously

3 210   3 200

evolution of methanogens
($CO_2 \rightarrow CH_4$)

assumed and quite similar to atmospheres of Jupiter and Saturn, though with less hydrogen, because hydrogen, the lightest of all gases, had already escaped into outer space due to Earth's relatively weak gravitation.

Enigma of the primitive Earth's atmosphere

We now know the compositions of Jupiter's and Saturn's atmospheres. These giants of the Solar System are composed almost entirely of hydrogen, which is mostly found in gaseous state, though some of it is liquid and some is in a strange form called metallic hydrogen which forms the cores of these planets. To be precise, Jupiter's atmosphere consists of 89.8% hydrogen and 10.2% helium by volume, with traces of methane (≈0.3%), ammonia (≈0.026%), water vapor (0.0004%), and minute amounts of ethane, hydrogen sulfide, neon, oxygen, phosphine, carbon, and sulfur. In fact, it is even considered to be a kind of miniature star, formed from the same elements as our Sun, but as its mass equals 1/1047 of that of the Sun, it is far too small to initiate a thermonuclear reaction. (The Sun is composed of 91.2% hydrogen and 8.7% helium to the total number of atoms.)

Saturn's mass is 95.1 times greater than that of Earth, and Jupiter is three times as massive as Saturn (it is exactly 318 Earth's mass). However because of their low densities, which are close to that of water, gravitational pull on Saturn is only 10 percent stronger than that on Earth, and we would not get flattened even on Jupiter, as popular wisdom holds – we'd weight only 2.5 times as much as on Earth. The temperatures of these giant planets are quite low (Jupiter's average is a chilly –145 ºC), and thus the velocity of hydrogen and helium molecules is very low as well. (The most probable velocity of a molecule is proportional to the square root of the ratio of gas temperature to molecular mass.) The consequence of all this is that molecules of helium and hydrogen are unable to reach escape velocities of their planets, i.e. 59.5 km/s on Jupiter, which would cause their degassing. On Earth, however, escape velocity is just 11.2 km/s, and it takes only a thousand years for hydrogen content in the atmosphere to be reduced to a third of its original quantity. (For this to happen with oxygen it would take 1026 years – much, much more than the age of the Universe!)

It is not difficult to calculate that just a few thousands of years, a mere blink on geological time scale, would be quite enough to deplete Earth of all of its original hydrogen. And so the early Earth's atmosphere, consisting of 10% hydrogen, 20% methane and 20% ammonia, which Stanley Miller reproduced in his experiments, might have been too rich in hydrogen. And what about water? Although water vapor as such was not added to the gas mixture, the mixture itself was saturated by it.

To sum it up, he made an apparatus which can now be found in countless textbooks, even in high-school ones. A big five-liter flask out of which protruded two tungsten electrodes was connected with the tube provided with a water cooler, leading further to a condenser (U-tube) and then to a small flask with 200 milliliters of boiling water and then again into the big flask with electrodes. "The water in the small flask is boiled to promote circulation and to bring water to the vicinity of the spark so that oxygenated organic compounds can be formed," Miller wrote. "The products of the discharge are condensed and flow into the flask through the U-tube, which prevents circulation of the water in wrong direction."

It is often claimed that Miller simulated the action of solar UV radiation on Earth's atmosphere. I am not really convinced of that. If you carefully read his 1955 paper, you will find that he knew very well that energy of ultraviolet radiation considerably exceeds that of lighting discharges. But on the other hand, it would be much more difficult to simulate the influence of radiation in a laboratory than to do the same for atmospheric electricity. The apparatus was made out of ordinary laboratory-grade Pyrex glass which is opaque to ultraviolet radiation (fluorite is the only material which is transparent to all UV radiation), and there were other technical difficulties as well. And finally, "it is possible that ultraviolet light would not be the principal source of organic compounds," wrote Stanley, because – according to him – it influences only the upper atmosphere, and atmospheric interactions with the ocean are crucial for the formation of life. How terribly wrong he was! Our planet would be sterilized entirely without its protective layer of ozone. Such a layer didn't exist in the primordial Earth's atmosphere, because there was no free oxygen in it.

Something pink in the flask

""During the run the water in the flask became noticeably pink after the first day, and by the end of the week the solution was deep red and turbid," Miller described his experiment in the first paper published in May 1953 issue of *Science*. (He asked of his supervisor to be listed as coauthor of the paper but Urey politely refused: "I've already got my Nobel Prize.") The experiment lasted one week and the doctor-to-be noticed that "most of the turbidity [of the final solution] was due to the colloidal silica from the glass," a note which provides a connection with another historical chemical experiment. It was the famous Lavoisier's "101-day experiment" by which "the father of chemistry" demonstrated that prolonged boiling of water does not transform it into earth, as was commonly believed, but that the "earth" so obtained is really the vessel glass degraded by hot water.

The rest was routine. Organic compounds were extracted with ether and then further separated by two-dimensional paper chromatography. Stanley put a few drops of the solution onto a sheet of paper similar to the one used for filtration. He then let the mixture of butanol and acetic acid flow in one direction and – as the paper dried out – a mixture of phenol and ethanol in the other, perpendicular to the first. After spraying it with ninhydrin (a specific reagent for amino acids), red spots appeared on the paper and from their positions it was possible to identify amino acids. He found glycine and alanine, the simplest of protein amino acids.

And so the work continued. After improving the separation procedure, Miller published the discovery of α-aminoisobutyric acid and sarcosine (N-methylglycine) in his 1955 paper and confirmed the findings of glycine, alanine, β-alanine, and α-aminobutyric acid. He also found the three simplest organic acids (formic, acetic, and propionic) and two simplest hydroxy acids (glycolic and lactic). He also found that 8.7% of added carbon (i.e. methane) converted into simple acids and 5.5% into amino acids. Too much or too little? It depends on your viewpoint; Miller's experiment lasted a single week while Mother Nature had eons at her disposal.

Not to be daunted, the young chemist made still further improvements to his apparatus. He modified the circulation system of his second device, and for the third one he used silent electric discharge instead of sparks. In addition, he analyzed the gases remaining after the experiment. Surprisingly or not, he found three new gases not originally present in the mixture. They were nitrogen, carbon monoxide and carbon dioxide.

But this was not the end of the story. In 1970 Stanley Miller and his collaborators repeated the experiment, now supported by a much better analytical method than paper chromatography. They found 33 different amino acids, including one half of the 20 commonly found in proteins. The popular

3 |20

formation of Vaalbara, the
first super-continent

Chemicals in Miller's experiment

Reactants (stuff that goes in)	Formula
Hydrogen	H_2
Methane	CH_4
Ammonia	NH_3
Water	H_2O
Products (stuff that gets out)	
Glycine (2-aminoethanoic acid)	H_2NCH_2COOH
Alanine (2-aminopropanoic acid)	$CH_3CH(NH_2)COOH$
α-aminobutyric acid (2-amobutanoic acid)	$CH_3CH_2CH(NH_2)COOH$
α-aminoisobutyric acid (2-amino-2-methylpropanoic acid)	$(CH_3)_2C(NH_2)COOH$
Sarcosine (*N*-methylglycine)	CH_3NHCH_2COOH
β-alanine (3-aminopropanoic acid)	$H_2NCH_2CH_2COOH$
Formic acid (methanoic acid)	$HCOOH$
Acetic acid (ethanoic acid)	CH_3COOH
Propionic acid (propanoic acid)	CH_3CH_2COOH
Glycolic acid (2-hydroxyethanoic acid)	$HOCH_2COOH$
Lactic acid (2-hydroxypropanoic acid)	$CH_3CH(OH)COOH$

image of Miller's apparatus began to resemble a *mastarion* (breast-shaped cup) of the alchemists, capable of creating life, just as mother's breasts supposedly are. An anonymous American chemist, a descendant of Jewish immigrants from Eastern Europe, turned into a hero of science, confirming the notion that a career in science cannot be planned. (In his later career he didn't manage to achieve anything important enough to be remembered by history.) But is all this really true?

Chemistry and antichemistry

From a chemist's perspective, Miller's undertaking is quite straightforward – he made an experiment on the uncontrolled radical reactions in the gas phase. Allow us to elaborate: gas molecules react with charged or non-

3 100   3 090

fossils of *Eobacterium isolatum*
(Central Transvaal, Africa)

charged fragments of molecules produced by electric sparking, silent electric discharge, ultraviolet or nuclear radiation, X-rays, heat, fire or whatever. These reactions may be quite complex, but they are not at all uncommon. Due to action of ultraviolet radiation, despite the protection of ozone layer, the air around us is constantly ionized, and it is – consequently – filled with all sorts of radicals.

Radicals from the air undergo chemical combinations with organic as well as inorganic molecules, even the ones in our bodies; during our lifetimes we are exposed to a cool fire kept burning by the very air we breathe. Methane released into the air is very quickly oxidized into water and carbon dioxide due to action of oxygen radicals. Our body also suffers this radical burden, for nearly half of the oxygen we breathe is not put to good use, to sustain our living processes, but for evil, to damage our protein and DNA molecules. There are natural biochemical mechanisms our body uses to get rid of radicals, and there is a range of pharmaceutical preparations which claim to be capable of reducing their amount and thus their damaging effects. Much research has been done on the reactions of radicals, but we still remain largely in the dark regarding that subject. This chemistry is so complex that it is extremely difficult to predict anything. But, to be fair, Stanley did not claim to predict anything, however he did succeed in proving what he set about to prove. In a way, his experiments belong in the murky field of *antichemistry*.

Why antichemistry and what is, after all, this antichemistry thing? Every chemist wants to obtain a single product in as high a yield as possible. And so he conducts his experiments in tightly controlled conditions; this and that chemical, heat to this or that temperature, mix and stir for five, ten, or twenty minutes. Use balance, pipette, burette and a stop-watch. But first of all, use known reagents and study the mechanism of reaction before you even think of commencing your experiment. All those sound principles of chemical science were missing from Miller's famous experiment. This makes me hark back to my childhood when I performed "experiments" by mixing all available chemicals, guided by a few reactions written in my textbook and led by pure chemical intuition. How I managed to survive to the end of my self-education in chemistry without any (serious) injury is still a mystery to me. But it is exactly this kind of chemistry which was practiced at one of the best American universities and under supervision of a Nobel Prize laureate!

Miller was neither the first nor the last chemist to achieve prebiotic synthesis, not to mention synthesis of amino acids. In 1912 Trier demonstrated

3 060

clearance of the oceans of Fe(II) ions
due to rise of free oxygen

that amino acids may be formed by reaction of ammonia with hydroxy acids, such as that of ammonia and glycolic acid producing glycine (this could also be the synthetic route for this amino acid in Miller's experiment). To be fair, in his 1955 paper Stanley did refer to his forerunners, namely Loeb's work from 1913, but in so doing he was far from modest: "The only work that would have any bearing on the reducing atmosphere would be the elements of Loeb who obtained glycine by the action of silent discharge on a mixture of carbon monoxide, ammonia and water."

Incidentally, while his first manuscript was still under review at *Science*, another manuscript by Kenneth Wilde and coworkers was following the same procedure in the same journal, describing electric arc synthesis from carbon dioxide and water which pointed to "implications with respect to the origin of living matter on earth." But this is just one piece of a much wider question of discoverers, co-discoverers and priorities. When in one of my popular science manuscripts I noted that the Montgolfier brothers invented the hot air balloon, the editor objected claiming that the first balloonist was actually some Portuguese gentleman, who was to me, then and now, a totally unknown person. If we start thinking this way, I complained, we'd have to base our discourse on the facts that America was not discovered by Columbus, but by Vikings or, more accurately, by Siberian tribes which crossed the land bridge which once connected Asian and American continents and in doing so became today's Native Americans.

The question of discoveries and scientific revolutions in general, is not that of who first saw something or had some new idea, but who made a significant impact on the scientific community and society. Haldane had ideas very similar to those of Oparin, but Oparin was lavishly supported by the "state of the working people," in contrast to Haldane who was not a very popular figure in Britain because of his Marxist convictions.

And Miller's experiment should be judged along these lines as well. It provoked an avalanche of similar experiments, similar research and similar papers. Everyone was hopping onto the wagon to find this *mixtura mirabilis* from which "live" molecules would finally emerge. This American scientist was a true counterpart to the Russian one; Oparin performed almost no experiments and Miller, with his followers, tried to experimentally explore each and every possibility. "Thus, the primary formation of compounds of the protein type is in no way unusual, exceptional, or different from the formation of other complex organic substances," wrote Aleksandr Ivanovich

3 030

clearance of the oceans of Fe(II) ions
due to rise of free oxygen

laconically, not bothering to support the statement experimentally at all. If general principles are known, everything is known, predictable and certain. Just like the Communist classless society. (Oparin even refused to accept Miller's results outright when he learned of them.)

But then again, nothing in science is certain. Every hypothesis can be proved or disproved, and every theory accepted or refuted. And asking the right questions is of utmost importance. What is not disputable is that a young American chemist managed to obtain amino acids, "the bricks of life," but did his gas mixture accurately represent the first atmosphere of Earth? And how did amino acids polymerize into proteins despite Oparin's claim that "the primary formation of compounds of the protein type is in no way unusual?"

Dreaming and waking up

The 1950s were the age of enthusiasm. After six long years of anxiety, death and misery which naturally accompany any war – blue, blue skies were here again... The world appeared as it had on the first day of creation, "and God said, 'Let there be light,' and there was light. And God saw that the light was good..." All those scientific achievements and technical innovations developed for the destruction of humanity could now be utilized for its well-being. New types of household devices, electric sewing and washing machines, electric pots, canned food, powdered milk and especially powdered eggs, popularly called "Truman's eggs" in Europe where they were a noted component of American postwar aid, were all miracles of a new age. It was an age in which everything appeared achievable, everything within reach of an outstretched hand. No one doubted that "in the future" at our disposal we'd have an inexhaustible source of clean energy in nuclear power plants, that we would spend our vacations in orbit around the Earth, on the Moon, or even on Mars. And the same was also held to be true with science.

The year 1953 was truly an *annus mirabilis*. In the very year Stanley Miller made his famous experiment, Watson and Crick solved the structure of DNA, the substance genes are made of, after eight years of research Frederick Sanger completed the amino-acid sequence of the first protein (insulin) and Max Ferdinand Perutz invented the heavy atom method in X-ray crystallography. In 1960 this enabled him and Kendrew to solve the first 3D structures of natural proteins, myoglobin (Kendrew) and hemoglobin (Perutz), and to share the 1962 Nobel Prize in chemistry. In 1954 Du Vigneaud

3 000

the first cyanobacteria (?)

synthesized the first protein (oxytocin) and in the same decade Linus Pauling found that protein molecules are arranged into helices. The next year "scientists created life," as was reported in the press, by rearranging tobacco mosaic virus (TMV) from its protein and RNA components (the "scientists" were Robley C. Williams and Heinz Fraenkel-Conrat). The mystery of life was solved once and for all; as was the mystery of its origin. Oparin provided the general theory, and Stanley Miller demonstrated by a simple yet convincing experiment that it works. Blue, blue, blue skies...

But it's time for us to wake up. If the Sun and the planets were formed from the same cloud of gas and dust, they had to share the same elements, and their original composition had to be the same. This was the prevalent thought at the time. Nowadays this is thoroughly refuted on the grounds of new data and computer simulations, but nevertheless let us posit a simple question – what was it that actually happened with Earth's original hydrogen and helium?

If early Earth was as cold as Jupiter, it would be difficult to imagine life emerging on a planet with liquid hydrogen ocean and liquid water being found only in the form of droplets swirling in the lowest cloud layer (the other three layers are composed of water ice, ammonium hydrogen sulfide, and ammonia crystals). If early Earth happened to be a warm planet with liquid water, its hydrogen would escape from its atmosphere quite rapidly. It was already mentioned that it would require no more than a thousand years to reduce hydrogen content of Earth's atmosphere to just one third of its initial value. A simple calculation shows that in just five thousand years its concentration would drop below 1% ($1/3^5 = 0.41\%$). Is it possible that life came into being in a just few thousands of years? Hardly.

The second problem has to do with other constituents of the first Earth atmosphere. "In this apparatus an attempt was made to duplicate a primitive atmosphere of the earth, and not to obtain the optimum conditions for the formation of amino acids," Stanley Miller stated clearly in his first paper. But is this really so? How could he possibly know that this primitive atmosphere contained equal amounts of methane and ammonia? Jupiter's atmosphere contains about 0.3% methane and 0.026% ammonia, by volume. This means that methane to ammonia ratio was closer to 10:1 than 1:1 as Stanley assumed. Degassing of planets is a process similar to distillation, and distillation is a marvelous process indeed because everything may be obtained from the same initial mixture, just like gasoline and kerosene are both produced from crude oil.

2 970

|

clearance of the oceans of Fe(II) ions
due to rise of free oxygen

But is distillation really a process so miraculous we can accept it as sufficient explanation of why primordial atmosphere of Earth consisted of equal parts ammonia and methane? Ammonia is quite soluble in water, cold water especially, and it is easy to imagine virtually all ammonia dissolving in the waters of the first ocean. Stanley knew this, but he didn't care. "The uncertainty of the relative amounts of methane and ammonia on the primitive Earth and the uncertainty of the size and temperature of the oceans which would dissolve much of the ammonia make any choice of relative amounts of gases just as arbitrary," he wrote in his 1955 paper. Then why did he put equal volumes of methane and ammonia into his apparatus?

The answer is quite simple. In his own words: "The methane and ammonia were taken in equal amounts so that appreciable quantities of carbon and nitrogen would react in the spark." In other words, if you want to prove that amino acids could have been produced in primitive Earth conditions, you have to prepare a gas mixture of such a composition to confirm your hypothesis! What kind of scientific argumentation is this?

The proper gas mixture

A proper mixture of gases seems to be the basic prerequisite for success of Miller's experiments. For example, a much better outcome was obtained with a mixture of methane, ammonia, carbon dioxide and hydrogen sulfide. Miller tried out such a mixture in 1958, but never got to analyze the products. Fortunately, his samples were preserved and after he passed away in 2007, his colleague Eric Parker analyzed these ancient samples and found 22 amino acids, 10 of which have never been found in any similar experiment. He also found amines, which had not been reported in the first two Miller's papers, and seven organosulfuric compounds, including amino acid methionine. "The results of this study may provide clues about the roles that primordial volcanic plumes may have played in the formation of some of Earth's first organic compounds," Parker wrote, pointing to the crucial influence of hydrogen sulfide. The chemistry of gases is a very complex one and a very small difference in gas composition can lead to high variance in types and quantities of produced organics.

But what was the actual composition of the very first Earth atmosphere? When ammonia was replaced with nitrogen, the modeled atmosphere was reducing as well, but when it was exposed to ultraviolet radiation only

2 940

the Moon is still very close to Earth;
tides 300 m high

alcohols and aldehydes were produced. No amino acid was detected. This was demonstrated in 1975 with an experiment by J. P. Ferris and C. T. Chen. Ammonia is obviously crucial to the production of amino acids, but in those days it was quite scarce in the atmosphere. As was said before, its presence in the primordial atmosphere should have been negligible due to its high water solubility. Even worse, ammonia on primitive Earth would have been destroyed by ultraviolet radiation.

Processes of this kind are well known on Venus where all water had been converted into hydrogen and oxygen through photochemical reactions a long time ago. Hydrogen escaped and free oxygen oxidized organics into carbon dioxide, turning "Earth's sister" into a vision of hell due to a "runaway greenhouse effect." In a similar manner, ammonia is converted into hydrogen and nitrogen. According to calculations, it takes a couple of hundreds of thousands of years to convert all atmospheric ammonia into its constituent elements. It is thus highly improbable that primitive Earth's atmosphere contained ammonia. If this is true, is Miller's experiment making any sense at all?

Evolution *vs.* creation

"Be prepared to talk about God," I said to my friend and his colleague, both professors of geology at the University of Zagreb, while enjoying our beer on a terrace facing the central building of the City Library where we've been invited to lead a discussion group on the origins of life. The first professor, my friend from student days, has been telling me that he indulged in beer for medicinal purposes only – as a remedy for his kidney stones, so whenever we met I'd always order "the beer this gentleman is drinking." But never mind the beer; I was invited to organize short lectures on the topic and then to lead a discussion with the public. And what kind of public? God knows who'll come. One could expect university professors, scientists, students, retired cooks and clerks, even schoolboys. And that's how it actually turned out.

My two colleagues spoke about conditions on the early Earth and on formation of planetary systems, while my topic was chemistry which might have led to the first living cell. And then came the questions. The first one was something about DNA and synthesis of nucleic acids, the second one was also about genes – but it was presented in a very unusual manner.

"You surely know me well," a tall slim man in his early forties addressed the beer-loving professor, adding "I used to visit your department." (Oh no!

He knows him!) And then he came to the point: "We can't have two truths – there can be only one. According to The Revelation, God created life and all the species that live on Earth. Man couldn't have evolved from animals for the one obvious reason – he is immortal. The original sin is transferred from generation to generation genetically..."

"Well, then why don't you find the original sin gene!" I exclaimed, although I immediately became aware of how harsh this sounded. In my position it was certainly quite inappropriate of me to make fun of anyone. I expected laughter, but nothing happened. As we concluded the session, the enthusiastic professor's acquaintance privately asked me how he could find that gene. "The human genome has been solved," said I, "so you can find what you need. Please, use Google." And he took me seriously.

But he was not alone. As I expected, the discussion turned from scientific to theological topics. Surely, I am not a theologian and neither are my two colleagues. How could I respond to this? I guess I couldn't do much better than Darwin two centuries ago in conclusion to his *The Origin of Species*: "Nevertheless they [his opponents] do not pretend to that they can define, or even conjecture, which are the created forms of life, and which are those produced by secondary laws. They admit variation as a vera causa in one case, they arbitrary reject it in another, without assigning any distinction in the two cases. The day will come when this will be given as a curious illustration of the blindness of preconceived opinion. These authors seem no more startled at a miraculous act of creation than at an ordinary birth."

But has the *day* come? Not yet. On *darwinismrefuted.com* one can read many arguments which seemingly disprove Darwin's theory. To be fair, every scientific theory, especially those in the fields of geology and paleontology, has its flaws. Nothing in science is definite because nothing in experience is definite. No theory is final because no human experience is final. You can doubt anything and everything, especially when it pleases you. Darwin's arguments should satisfy every scientist as well as every pious man; why would God abandon natural laws only in those cases where scientists haven't yet arrived at an appropriate theory? It seems to me that it is simply much easier to attribute origin of life to a miraculous act of creation than to study scientific facts.

It is futile to use Miller's experiment to criticize Darwinism. "Although nearly half a century has passed, and great technological advances have been

2 880

clearance of the oceans of Fe(II) ions
due to rise of free oxygen

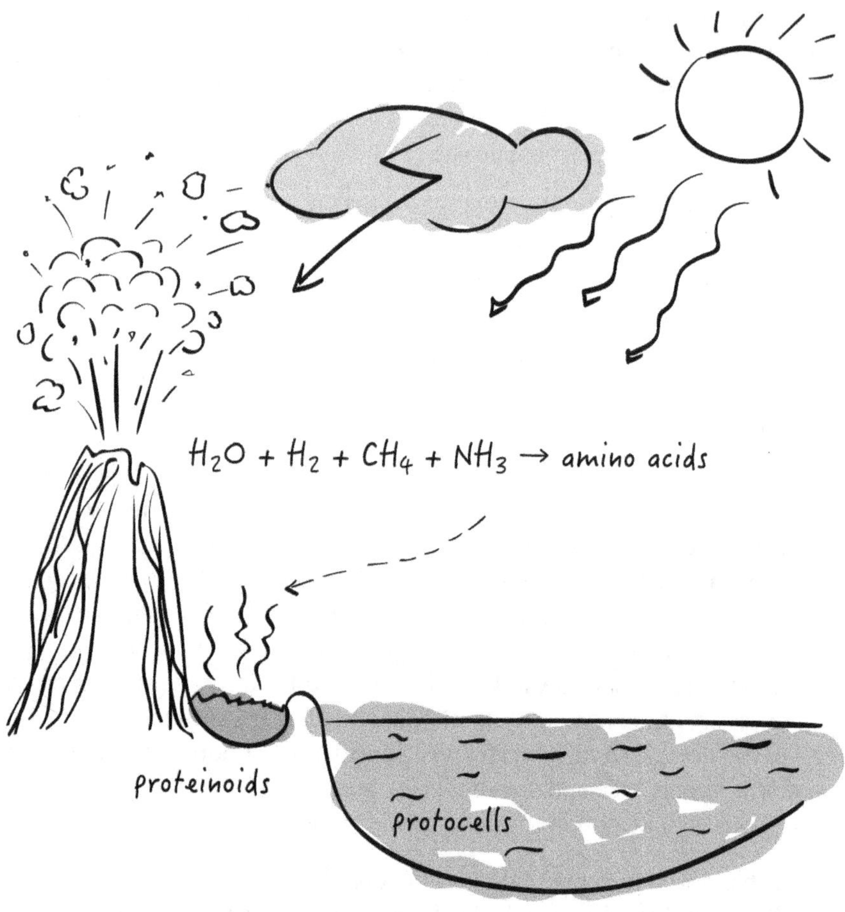

There are many ways to convert simple organics into amino acids. Their subsequent polymerization very probably occurred on the slopes of volcanic mountains which provided heat for the formation of proteinoids, ancestors of all proteins.

made, nobody has made any further progress," wrote an anonymous author on the *darwinismrefuted.com* page. "In spite of this, Miller's experiment is still taught in textbooks as the evolutionary explanation of the earliest generation of living things. That is because, aware of the fact that such studies do not support, but rather actually refute, their thesis, evolutionist researchers deliberately avoid embarking on such experiments."

2 850

clearance of the oceans of Fe(II) ions
due to rise of free oxygen

But therein lies the root of the misunderstanding. Miller's experiment is not a new revelation, miraculous discovery of some hidden truth, but simply a bit of evidence. It is an answer to the question of possibility of "production of amino acids under possible primitive Earth conditions," as was suggested by the very title of his first 1953 paper. And as with all scientific works, it creates more questions than answers. The first question is did the "primitive Earth conditions" look like those Urey and Miller arrived at? This question has already been answered. The second question is what happened later with amino acids dissolved in primordial Earth's ocean? Life cannot be formed from free amino acids. They have to bind together into protein molecules first.

The emergence of proteins

Proteins are composed of amino acids even though they have little in common with them. To put it plainly, proteins are amino acid polymers (polyamino acids) and the relation of proteins to amino acids is chemically equivalent to the relation of glucose to cellulose or ethylene gas (ethene) to polyethylene. Proteins may decompose into amino acids through heating with concentrated hydrochloric acid, or by action of enzymes, which is what happens in our stomach and duodenum. On the other hand, achieving the reverse and binding amino acids together into proteins is a much more difficult task.

In order to form a peptide bond ($-CO-NH-$) it is necessary to bind a carboxyl group ($-COOH$) of one molecule to an amino group ($-NH_2$) of another. To that end chemists use reagents which protect the amino group of the first amino acid and activate the carboxyl group of the second one, or *vice versa*. In living cells the process is governed by tRNA molecules. The first tRNA molecule is bound to the carboxyl group of incoming amino acids and the second one to the carboxyl group of the growing polypeptide chain. In this way terminal carboxylic group of the peptide chain is prepared for the action of the amino group of the new amino acid that has to be incorporated into it. It is a rather complex process as it involves tRNA, mRNA, ribosomes and a multitude of enzymes. It is difficult to even imagine that first proteins were synthesized in this fashion.

But then again, there is another way. As mentioned before, said proteins can be decomposed into amino acids by action of acids. It is a hydrolytic reaction; it is water that is actually doing the job of breaking the peptide bond ($-CO-HN- + H_2O \rightarrow -COOH + H_2N-$), and not the acid. However, every

2 820

clearance of the oceans of Fe(II) ions
due to rise of free oxygen

chemical reaction ($\rightarrow$) may be reverted ($\leftarrow$). Therefore, amino acids can be polymerized into peptides by release of one water molecule per peptide bond ($-COOH + H_2N- \rightarrow -CO-HN- + H_2O$). Is it possible that this reaction could have occurred on the primitive Earth?

Five years after Miller's experiment Sidney W. Fox and his coworkers from Florida State University published papers on "thermal copolymerization of amino acids to a product resembling protein." They obtained their polymers by heating (above the boiling point of water) and then drying mixtures of amino acids. These polymers, named proteinoids, were capable of forming "microspheres," resembling Oparin's coacervate droplets and have even demonstrated catalytic activity, resembling enzymes. But there was a problem. If the mixture of amino acids was heated for one hour at 160 to 190 ^{o}C, in addition to amino acid polymers other products were obtained as well. This is not at all surprising because amino acids are not very stable, and they easily decompose into carboxylic acids and may even convert to aromatic compounds.

To a simple chemist's mind it is difficult to comprehend that it took twenty years to solve this problem. In 1976 Duane L. Rohlfing performed the same experiment as Fox did, but at lower temperatures. He found that polymerization of amino acids is possible even at temperatures as low as 65 ^{o}C, though his experiment lasted for 81 days.

"Time, however, was a plentiful prebiotical commodity," he noticed, pointing to a ubiquitous phenomenon taught at the introductory chemistry course, and that is that the rate of chemical reaction is doubled or even tripled by each ten degrees of temperature increase. This means that polymerization of amino acids may occur at *any* temperature, if proper drying is provided, of course. Moreover, exact calculations have shown that at 35 ^{o}C the mixture of amino acids would yield the same extent of polymerization as observed after 81 days at 65 ^{o}C, if one is patient enough to wait for seven years, that is.

Seven years, seven hundred years, seven hundred thousand years, and more... Many hundreds of millions of years ago, many hundreds of millions of years passed away... And in my vision I saw the "warm little pond" full of amino acids and other organics. The pond was dry, devoid of water long evaporated by hot rays of the Sun. And then the clouds came and the pond filled with water. And then again came the drought... Followed by rain and then drought and again, and again, and again... There were many such "warm little ponds," many dry days and many rainy days, and thus, as Rohlfing said, "the formation of polyamino acids may have been a relatively widespread and common event on the prebiotic Earth."

But what kind of molecules were they? The first polyamino acids were random polymers, a chaotic combination of all sorts of amino acids. How on earth could anything like an ordered protein molecule emerge from that stuff?

But this is science: perhaps you can offer a better hypothesis?

The moral of Miller's experiment

Columbus discovered America. No, he didn't. Columbus never actually reached the American continent. Instead, he landed on Caribbean islands convinced that he arrived at *Cipang* (Japan). Galileo discovered Saturn's rings, Jupiter's satellites, and mountains on the Moon. No, he didn't. Nobody knows what he actually saw through his *telescopium*. His instrument was designed for his eyes only, and nobody except those who by coincidence were as farsighted as Galileo could see anything. I said "anything," because chromatic aberration was so severe that it really took an extreme effort to see anything. Democritus was the father of atomistic theory. No, he wasn't. He had just put forward a fuzzy hypothesis, an absurd notion, which in no way could be proved. And so on.

I remember an old book from my boyhood titled *A History of Delusions* describing all kinds of absurd and ridiculous scientific theories. But we are talking about the history of science here! We now know that there are no channels on Mars, no intelligent beings on the Sun, but we do know that Mars has a landscape and that black spots on the Sun's surface are causing aurora borealis and regularly menacing our electronics and electric systems (rather than being holes in the Sun's fire clouds, as Haeckel believed). Great scientific discoveries are not those which once and for ever establish undeniable truths, but those that ask new questions and point to new directions. And that's exactly how it is with Miller's experiment.

If we took his experiment at face value, that he simulated processes in a Jupiter-like atmosphere, we would come to an obvious conclusion that there is life on Jupiter and that there isn't one on Earth. But if Miller's gas mixture was actually composed of volcanic fumes mixed with primitive Earth's atmosphere, as his forgotten experiment suggests, then we come much closer to a solution.

2 760

clearance of the oceans of Fe(II) ions
due to rise of free oxygen

And finally, it seems more plausible that life emerged on the slopes of volcanic mountains than in waters of an ocean. Or maybe it emerged on the slopes of a volcanic mountain near an ocean, on its coast. Maybe.

"Maybe" and "perhaps" are legitimate terms in science, but only if they lead to a proof. "Science is making of the something what might be to something what is," pronounced Hungarian chemist Albert Szent-Györgyi, discoverer of vitamin C. Perhaps God did create the world, but this *hypothesis* has no scientific relevance. Could we ever know how God created the world? Could we propose any experiment starting from this hypothesis? Could we reveal any secret of nature by assuming it to be a creation of God's? No. "To anyone who is not presumptive there is only one solution: life originates only from life; once there was no life on Earth; now it is here; thus it is the creation of a supernatural being, who is the highest form of life, the fullness of life, and is called God." This is the last sentence in a popular book written by a Franciscan monk and published 1932 in Croatia. What kind of science could emerge from this? Probably not even a proper religion.

On the other hand, materialism may also be considered a kind of religion. Was life created by the supreme will of the Supreme Being, as revealed to Moses and Jesus, or did it slowly develop according to the laws of dialectical materialism, as claimed by Marx and Engels? "Engels shows that a consistent materialistic philosophy can follow only a single path in the attempt to solve the problem of the origin of life," wrote Oparin in his famous book *The Origin of Life*. I maintain that Engels doesn't show us anything. He only expressed belief in his materialistic philosophy. But Oparin does show something; to some extent at least.

Oparin did not present any convincing proof in favor of his hypothesis. But he showed us the way. The problem of the origin of life may be solved if there is enough experimental evidence supported by sound hypotheses and theories, geological as well as paleontological findings, astrophysical data and cosmological insights. The first steps were chemical experiments aiming to simulate conditions on the primitive Earth. Miller was not the first one to perform such experiments, but his experiments did open up new possibilities. And for us, this is essential.

And there is more. He confirmed, or let us say – gave a new proof – of Oparin's theory thus enabling scientists to shape the general theory of the origin of life. According to this theory, life came into being in five steps.

2 730

formation of
super-continent
Kenorland

Five steps to life

The first step is formation of planets and planetary atmospheres. This produced a mass of relatively simple components – gases of a primitive, presumably reducing atmosphere.*

The second step is formation of simple organic compounds in the atmosphere. For this process some kind of external energy source is needed. Whether the source was lightning, ultraviolet or nuclear radiation is quite immaterial; what is important is that the atmosphere remained constantly ionized. This provided chemical agents in the form of highly reactive ions and radicals.

The third step required liquid water. It conserved the products of initial atmospheric reactions, which would have otherwise decomposed by action of the same chemical agents or ionizing radiation, which had created them. In this respect, water in the U-tube and in the little flask of Miller's apparatus is of utmost importance. Without those "water traps" we could expect very few products.

The fourth step led to further chemical modifications of the organic matter. According to Miller's view, as well as views of his followers, the key event was polymerization of amino acids into protein-like substances (proteinoids). This reaction requires nothing but heat, and heat is the most widespread form of energy.

The fifth and the last step is formation of the first protocells from polymeric organic material. Presumably, they were Oparin's droplets formed by coacervatisation of original proteinoids.

So here we have a consistent, though still not complete, theory of how life came into being. It is, no doubt, a protein theory. The origin of life is seen as an evolution of proteins, obviously because in those days nucleic acids were not yet perceived as important for life as they are now. Anyway, Oparin, Urey and Miller did their job.

And what is the final moral of Miller's story? Perhaps incidentally you did order root beer, but its repulsive taste inspired you to try finding a better beer during your stay in America.

*There is much discussion on the composition of the primordial Earth atmosphere. It seems that it originally consisted of methane, hydrogen and ammonia, and was later destroyed by the collision that formed the Moon. The new atmosphere which replaced it was composed primarily of carbon dioxide and monoxide.

---

2 700

traces of eukaryotes
in Australian cherts

2 670

Mill. yrs from present

maximum of the precipitation
of oxidized Fe(II) ions,
deposited as hematite (Fe_2O_3)

5. BONE SOUP

Perhaps life originated with the emergence of nucleic acids rather than proteins. A very interesting hypothesis – but does it fit the facts? (And what are those nucleic acids anyway?)

Take one pound (half a kilo) of large cut beef bone, thigh-bone for instance, place it into cold water and bring the heat up to the boiling point. As the meat begins to change color add some onions and carrots, and then salt and pepper. And that is the beef bone soup.

But there is another, much better version of the same soup. It is made from bones which were roasted previously, those that remain after the meat on them was eaten. This recipe comes from my mother who learned to cook during World War II, when most of occupied Europe experienced famine and there was no free market in food to speak of. In 1942 my mother's family, all five of them, were eligible to buy exactly 750 grams of bread (or 550 grams of flour) a day, and not a single gram more. Sugar, oil, and milk were all rare commodities, not to mention such incredible luxuries as meat and butter. Everything had to be eaten and nothing could be wasted. A soup was usually prepared from anything, even the tasty, nutrient-rich hot dishwater was used with relish. An army officer on the eastern front, a member of the Croatian contingent in the *Wehrmacht*, described in one of his letters home how he was regularly served common European breakfasts (coffee, milk, white bread, butter and marmalade). This was published in the newspapers and was read like a fairy tale. Soldiers had to eat, civilians could as well starve.

Civilians may have starved, but at least they were not being killed. Necessity is the best teacher, as a Latin proverb claims, and so a black market quickly established itself. New recipes emerged, and some persist even today. In one cozy Dalmatian restaurant I ordered a "shark in algae sauce." Algae sauce? Back then poor inhabitants of the Adriatic coast had nothing to eat, and they came to an idea to include sea grass – algae encrusted

2 640

clearance of the oceans of Fe(II) ions
due to rise of free oxygen

stone from the shallows – in their pots. A stone soup, essentially. And today it is enjoyed as a specialty.

And so are the bones. The smaller fish, especially when roasted or grilled, are eaten along with their bones – in one bite. And how small should a "small" fish be? It depends. To me the size limit is five centimeters, but there are people who eat bones from much larger fish regularly. The following scene took place in a fish restaurant. A visitor noticed another guest eating fish first, and when he finished, he ate their bones as well. Intrigued, he ventured a question: "Excuse me, but why are you doing it?"

– Doing it? What?

– You are eating fish bones!

– You don't know?

– What don't I know?

– Don't you know that fish bones are very rich in phosphorus? And phosphorus is very good for your brain. Listen, I'll let you have some of mine... for a small price.

And very quickly, the deal was done – fifty cents a bone. After a while the buyer, munching on his fish bones, realized: "Wait a minute! Half a dollar for one fishbone, but you can get a whole fish for only a quarter!" "You see? It works, it works!" exclaimed the gleeful con-man.

It may be that phosphorus is good for your brain, because its cellular membranes contain special fatty substances called phospholipids, but is there really any phosphorus to be found in a meal prepared from fish or beef bones? It is a very important question, actually – so important that the beginning of life on this planet hinges on it.

What is a bone?

"Bone is a rigid organ that constitutes part of vertebral skeleton," asserts the lexicon definition. Never forget that bone is an organ, that it consists of many tissues, and that it grows, develops, degrades and perishes like any other organ. Dead bone however, the one from which a soup may be prepared is something entirely different.

To be brief, the bone used in our soup consists of 30% organic and 70% inorganic matter. This organic matter is mostly collagen, while the

2 610

clearance of the oceans of Fe(II) ions
due to rise of free oxygen

inorganic part consists mostly of various salts, predominantly hydroxyapatite. Those two substances, collagen and hydroxyapatite are responsible for the strength of our bones. They are mutually interconnected and interwoven in a way similar to how bricks and mortar cling to each other to form a brick wall.

This dual chemical nature of our bones can be verified by two well-known experiments. The first one is to put a piece of bone in an acid. Hydroxyapatite will dissolve and collagen will remain. Or we can put bone into fire: collagen will perish while hydroxyapatite will remain unchanged. But what about our soup? What is happening in there?

In its essence, cooking is nothing more than denaturation of proteins. This means that water and heat change the shape (but not chemical structure!) of polypeptide chains in protein molecules. As collagen is a highly structured protein, cooking degrades it into random coil state. Therefore, cooking is a process of curling protein molecules, transforming them from a structured into an unstructured state. But what about their mineral component, hydroxyapatite?

The answer to this question can be readily found in our soup. After cooking, a bone looks remarkably like a stone. Hydroxyapatite did not dissolve, but it was separated from organic matter.

Calcium salts of phosphoric acid are not soluble in water. Hydroxyapatite is one such salt. It consists of calcium cations and phosphate and hydroxide anions; its chemical formula is $Ca_5(OH)(PO_4)_3$. Its OH component may react with acids, especially with hydrofluoric acid. In this case, hydroxyapatite turns into fluorapatite; the substance tooth enamel is made of. This reaction takes place in soil too, and enables us to date fossil bones from their fluorine content. By this process, inorganic matter in bones turns into a real mineral known as apatitee.

Apatite is usually pale green in color, but it can also be white, colorless, bluish, reddish, brown, grey, yellow or purple. It is a mineral found in igneous rocks and metamorphic limestones. The variety of its colors reflects the variety of its chemical composition. Its formula $Ca_5(F,Cl)(PO_4)_3$ defines apatite as calcium phosphate with a variable content of fluoride and chloride. It is, of course, insoluble in water, but it is soluble in hydrochloric and sulfuric acid – the fact widely used in production of phosphate fertilizers.

2 580

clearance of the oceans of Fe(II) ions
due to rise of free oxygen

Whether hydroxyapatite can dissolve in stomach acid is the crucial question in our story about fish-bone eater; if this cannot happen then there is no chance for phosphorus to benefit his brain. Solubility of apatite may also be crucial for the origin of life. How so? Well, it's a rather long story.

The stuff genes are made of

In 1953, the very year Stanley Miller made his famous experiment on the synthesis of amino acids by simulating chemical interactions in the primitive Earth atmosphere and its ocean, three papers appeared in the April issue of *Nature* journal under the general title "Molecular Structure of Nucleic Acids." The first one, titled "A Structure for Deoxyribose Nucleic Acid" begins with a statement of the type very rarely found in scientific journals: "We wish to suggest a structure for the salt of deoxyribose nucleic acid (D.N.A.). This structure has novel features which are of considerable biological interest."

The structure proposed by American biologist James D. Watson and his British colleague Francis Crick was the famous "double helix," the most fascinating and perhaps the most popular molecule in the world, besides H_2O, of course. Its structure is very simple. "The two ribbons symbolize the two phosphate-sugar chains, and the horizontal rods the pairs of bases holding the chains together," states the caption, adding "this figure is purely diagrammatic." "It is simply two spiral stairways joined together," I explained the structure to an architect who came to me with an idea to build a model of DNA molecule for the Museum of Evolution in the little town of Krapina, near Zagreb, known for the discovery of fossilized Neanderthal remains at the end of the nineteenth century. (It is interesting to note that 1–4% genes of modern humans living outside Africa can be traced to this ancient human race.) "Hand-holds are phosphate-sugar chains and steps are nucleic bases," I continued. This is all quite evident to a chemist, but to laymen not accustomed to chemical formulas it may prove to be something well beyond their grasp.

If you go along one of its helices, DNA molecule could be described by a simple formula –P–S(N)–P–S(N)–P–S(N)– and so on, where P stands for phosphate, S for sugar (deoxyribose) and N for one of its four nucleobases. Two of them, adenine (A) and guanine (G) are derived from purine, and are

2 550

---

clearance of the oceans of Fe(II) ions
due to rise of free oxygen

Nomenclature of the constituents of nucleic acids

Nucleobase	Type	Nucleoside	Nucleotide
Adenine (A)	Purine	Adenosine	Adenosine-5'-phosphate (Deoxy)adenylic acid
Guanine (G)	Purine	Guanosine	Guanosine-5'-phosphate (Deoxy)guanosylic acid
Uracil (U)	Pyrimidine	Uridine	Uridine-5'-phosphate Uridylic acid
Thymine (T) 5-methyluracil	Pyrimidine	Thymidine	Thymidine-5'-phosphate Deoxythymidylic acid
Cytosine (C)	Pyrimidine	Cytidine	Cytidine-5'-phosphate (Deoxy)cytidylic acid

therefore called purine bases. The other two, thymine (T) and cytosine (C) are called pyrimidine bases because they are regarded as derivatives of pyrimidine. Purine, as well as bases derived from it, has planar molecules composed of two fused rings, a five- and a six-membered one. Pyrimidine bases are a bit less complex for they are composed of only one, also planar, six-membered ring. Their basicity stems from nitrogen atoms in their rings. In other words, nucleobases are aromatic heterocyclic compounds, as chemists like to say.

"The novel feature of the structure is the manner in which the two chains are held together by the purine and pyrimidine bases," Watson and Crick described the structure of DNA. "The planes of the bases are perpendicular to the fibre axis. They are joined together in pairs, a single base from one chain being hydrogen-bonded to a single base from the other chain, so that the two lie side by side with identical z-co-ordinates." In short, the famous A-T and C-G pairing takes place, and "It has not escaped

2 520

clearance of the oceans of Fe(II) ions
due to rise of free oxygen

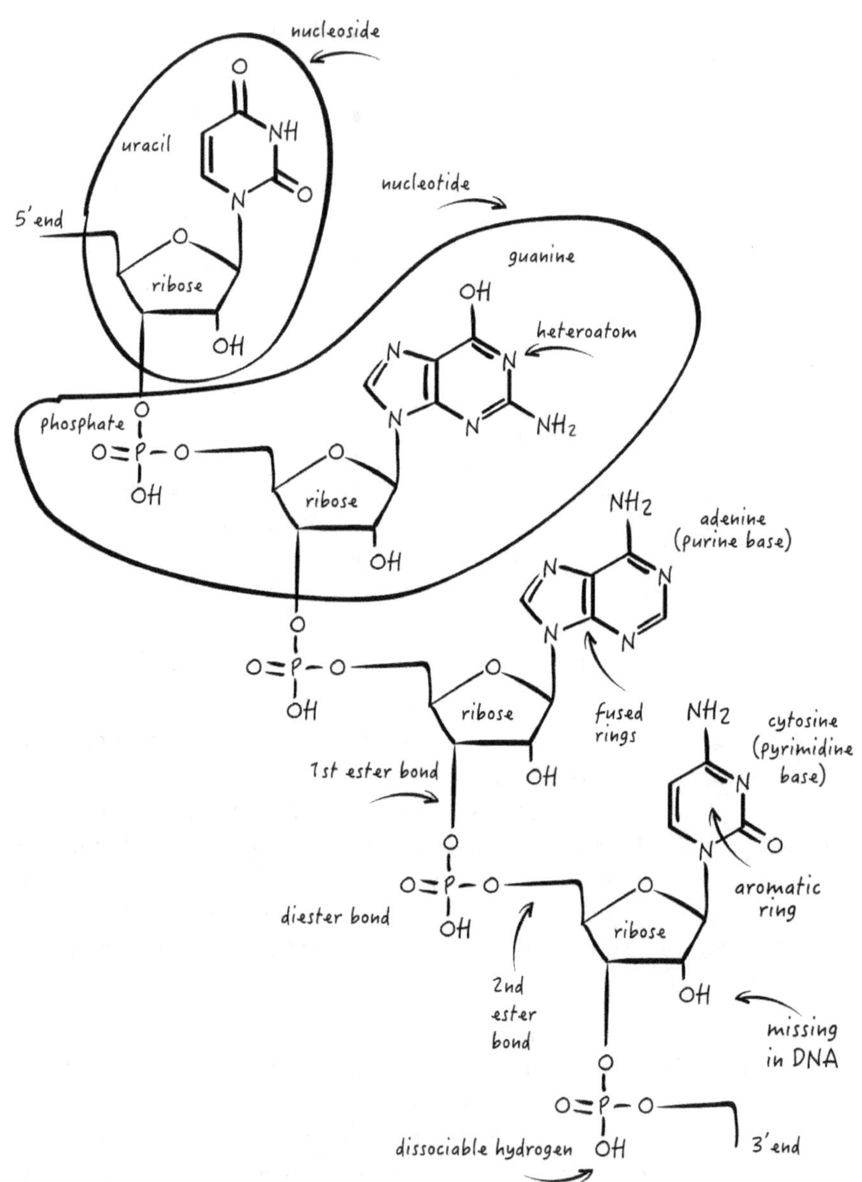

nucleoside
uracil
5′ end
ribose
OH
nucleotide
guanine
OH
heteroatom
N
N
N
NH₂
phosphate
O=P—O
OH
O
ribose
OH
O=P—O
OH
O
ribose
adenine
(purine base)
NH₂
N
N
N
N
fused rings
1st ester bond
OH
diester bond
O=P—O
OH
2nd ester bond
NH₂
cytosine
(pyrimidine base)
N
O
aromatic ring
O
ribose
OH
missing in DNA
O=P—O
OH
dissociable hydrogen
3′ end

2 500
2 490
Kenorland broke up

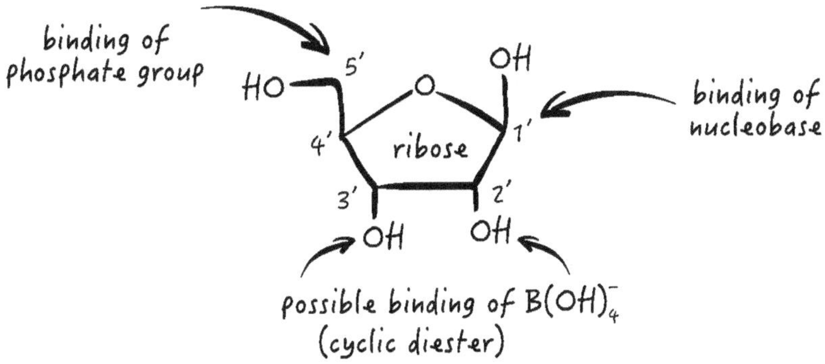

The structure of UGAC segment of ribonucleic acid (RNA). It may be considered as a polymer of nucleotides (polynucleotide) or, equally valid, as a diester (double ester) of phosphoric acid, H_3PO_4. However, nucleic acids in the cell are not acids but salts due to dissociation of "esterified" phosphoric acid: $(RO)_2PO(OH) \rightarrow (RO)_2POO^- + H^+$.

our notice that the specific pairing we have postulated immediately suggests a possible copying mechanism of the genetic material."

Is this "the formula of life?" Yes, certainly. But nobody expected it would turn out to be so unexpectedly clear and simple. It was quite well known that chromosomes are responsible for the transfer of hereditary information, but what exactly are those "chromosomes?" They are tiny particles composed of "nucleoproteins." This simply means that the substance of chromosomes may be broken into proteins (histones) and nucleic acids, that is DNA, and that further on into nucleotides, nucleosides, purine and pyrimidine bases, deoxyribose (a kind of simple sugar), and finally phosphoric acid.

"Now you can see from all the evidence that molecules of nucleoproteins are of a very complicated composition," wrote Croatian biochemist Fran Bubanović in 1918. "That is not at all surprising if we bear in mind that from these bodies the cell nuclei of spermatozoids are built, and we know that a single spermatozoid, that is hardly visible by the strongest of microscopes, transfers physical and psychological features from father to his son.

How complicated the structure of the molecule has to be to hold such a great power and perform such a great task!"

But the esteemed biochemist, as well as many other scientists of his time, was in error. The structure of genes is not complicated. It is very simple. "Aperiodic crystal or solid," as Austrian physicist Erwin Schrödinger, author of the famous wave equation, named the "hereditary substance" in his book *What is life?*, first published in 1944, was not a protein, as "all" scientists believed, but nucleic acid.

The way of all living

Structure of DNA may be simple, but it is not so for the structure of genes, and especially for mechanisms of their expression. Each gene encodes the sequence of the corresponding protein by a universal genetic code. It is based on the sequence of three consecutive nucleotides in a DNA molecule. Every three nucleotides encode one amino-acid residue to be incorporated into the polypeptide chain; GCA is the code for alanine (Ala), GGG for glycine (Gly), CCG for proline (Pro), CAC for histidine (His), and so on. The DNA sequence GCAGGGCCGCAC produces the peptide sequence Ala-Gly-Pro-His to give an example. This is a simple scheme, but the process is complex.

First the segments of DNA molecule, genes, have to be copied to smaller RNA molecules. They are molecules of one-stranded messenger ribonucleic acids (mRNA). Molecules of mRNA are then bound to ribosomes, complex bodies of about 15 nm across, and thus cannot be seen through a visible light microscope. Ribosomes are composed of proteins and nucleic acids, ribosomal RNA or rRNA.

The third key component in protein synthesis is my favorite – as is my mother's bone soup. Transfer ribonucleic acid (tRNA) was the topic of my BSc thesis in which I described how I managed to succeed in isolating and purifying 0.1 grams of tyrosine tRNA (tRNA^{Tyr}) from baker's yeast. It is a very interesting substance with structure in many ways resembling that of proteins. Cloverleaf folded and relatively small molecules (about 80 nucleotides long) of tRNA have two major functions. First one is to recognize the sequence of three nucleotides, also known as a codon, on mRNA molecules. This recognition is achieved by base pairing; tRNA for alanine (tRNA^{Ala}) has CGU (anticodon) in a particular position of its molecule sequence which is to be paired with the GCA sequence, the codon on mRNA molecule.

(Note that uracil, U, in RNA corresponds to thymine, T, in DNA.)

The key to "transferring physical and psychological features from father to his son" is in the way this recognition works. Nucleic acids recognize each other by base pairing, A-T and C-G for DNA and A-U, C-G for RNA. But there is a much more complicated recognition process that takes place between nucleic acids and enzymes, which are proteins. An enzyme has to recognize tRNA^{Ala} molecule and bind to it an alanine molecule to obtain Ala-tRNA^{Ala}, and not Gly-tRNA^{Ala} or Tyr-tRNA^{Ala}. Consequently, tRNA^{Gly} has to exclusively bind glycine (Gly) while "my" tRNA^{Tyr} has to bind tyrosine (Tyr), and so on. Each tRNA has to bind its own amino acid, and for every binding there is an enzyme to join tRNA and its amino acid together.

This recognition process is not simple at all. tRNA I had isolated and purified for my BSc thesis was used by my senior colleagues to elucidate the process of protein/RNA recognition, to find the "secondary codon" of tRNA molecule. In 1970s it was believed that enzymes recognize their tRNA's by a particular sequence of nucleotides. To put it simply, as a sequence of three nucleotides (anticodon) recognizes corresponding sequence on mRNA (codon), so the sequence of so-and-so nucleotides on tRNA (secondary codon) should recognize characteristic arrangement of amino acids of an enzyme.

But this turned out to be wrong. "There is no specific affinity between the side group of many amino acids and the purine and pyrimidine bases found in RNA," wrote James D. Watson in his *Molecular Biology of the Gene*, a textbook (published in 1970) which I read in two days and found as suspenseful as a detective story (No, really!). Many experiments have been published, many modifications of tRNA molecules have been performed, but they were all in vain. Nothing was found! Many years later I learned that the problem of "secondary codon" was ultimately solved in a way that was quite unexpected. It was necessary to solve 3D structure of tRNA molecules to find that an enzyme recognizes its tRNA by the general shape of the molecule. Each enzyme has a groove into which only a particular molecule of tRNA as well as the corresponding amino acid could fit. But where on Earth could a mechanism as complex as this originate?

All this didn't bother either Oparin or Urey and his ardent student Stanley Miller. They simply didn't know. They didn't have the slightest notion of the enormous complexity of gene expression mechanism. "From the viewpoint developed in the preceding chapters it seems much more probable that the

2 400

ice age (Huronian glaciation)

origin of the isolated nucleus, chondriosome, plastid, etc., is only the external visible expression of a gradual unfolding and perfection of an inner physico-chemical structure and organization of colloidal formation," was Oparin's view on the subject. "The very ephemeral initial orientation of molecules could acquire by an evolutionary process a more stable character and serve as the starting point for the formation of complexes and structures more or less easily distinguishable under the microscope." And that was all. Self-organization of proteins in their colloid form led to the first living thing. Who cares about nucleoproteins, not to mention nucleic acids?

But everything changed in 1953. Nucleic acids suddenly came to the fore. The mystery of life's origin cannot be solved without proposing a plausible hypothesis on how DNA and all that complicated stuff for its expression happened to emerge. Life is not a product of "living protein" but of a "selfish gene," a DNA molecule which reproduces itself from generation to generation, and in which all past, recent and future features of an organism are reflected. Evolution is, in its essence, a history of mutations, that is a history of nucleotide interchanges in DNA molecules.

But the question which has to be answered first is this: How did nucleic acids form in the primordial soup? Was it a bone soup? Perhaps.

Quest for the first ribose

"Did you hear that they found nucleic acids in meteorites?" a friend asked me expecting I'd be impressed. But he didn't succeed. I don't know if he heard it wrong, or was the mistake made by the journalist who wrote this article on "nucleic acids from space," but I do know that this did not and could not happen. What probably happened is that someone found nucleobases, or more probably heterocyclic aromatic compounds resembling them in meteorites. It is not at all improbable because virtually every organic substance which is heated, by a process known as pyrolysis or destructive distillation, gradually transforms into heterocycles. (This was known, in some way or the other, even by alchemists who prepared "castor oil," that is pyridine bases, by distillation of eggs.)

If we wanted to be more rigorous, we would describe pyrolytic reactions as cyclisations by condensation followed by thermal eliminations. In the first step, condensation, water vapor is released and in the second, known as thermal eliminations, carbon dioxide and ammonia are developed as well. The net effect is that organic substance becomes enriched with carbon by forming cyclic structures and new double bonds. "When it's smoking it's cooking, when it's burning it's done..." Never forget the ladle and how to use it!

The problem with abiogenic synthesis of nucleic acids is that even though they are polymers, as proteins are, they are not simple polymers of simple monomers, as proteins are simple polymers of amino acids. Nucleic acids have three components, nucleic bases, sugar (ribose or deoxyribose) and phosphoric acid. These components have to be first synthesized and then joined together in a particular way. Two molecules, that of nucleic base and that of phosphoric acid, have to find the third one, that of sugar, in order to form a nucleotide, the monomer of nucleic acid.

But there is another way of appreciating chemistry of nucleic acids. Following polymerization, a molecule of phosphoric acid binds two sugar moieties. One hydroxyl group of each sugar molecule is involved and two ester bonds (P–O–C) are formed. Nucleic acids are therefore double esters (diesters) of phosphoric acid. But phosphoric acid, H_3PO_4, has three hydrogens, so the third one remains free to give nucleic acids their acidity. Nucleic acids are, in effect, a special kind of phosphoric acid.

But from another perspective, from the viewpoint of a polymer chemist, nucleic acids are copolymers of nucleotides. Consequently, they can also be seen as polynucleotides. Since nucleotides are nucleoside phosphates, nucleic acids could be named, in quite a cumbersome fashion, poly(nucleoside phosphate). Their general formula is $(-R-O-PO(OH)-O-)_n$, where R stands for nucleoside, i.e. sugar-base moiety. Or maybe it would be even better to write down their formula as $(-R-O-PO(O-)-O-)_n$, because nucleic acids are actually ionized, existing in living cells in the form of magnesium salts.

Simple sugars may be synthesized, as mentioned before, by the formose reaction. If the simplest aldehyde, formaldehyde (methanal) were to be dissolved in a basic solution of calcium hydroxide, it would be transformed into sugar-like substances. This has been well known since the nineteenth century. But there are many simple sugars, monosaccharides; how could ribose be exclusively or at least predominantly formed?

The answer can be found in a vial of eye drops. It contains boric acid, a well-known germ-killer. The acid, H_3BO_3 or $B(OH)_3$, is similar to phosphoric acid, H_3PO_4, because both acids are tribasic – they have three dissociable

ice age (Huronian glaciation)

hydrogen atoms. But there is a difference. Boric acid exhibits some very interesting behaviors in alkaline solutions. At pH = 9.1 it binds an additional hydroxyl group, OH^-, turning itself into borate ion, $B(OH)_4^-$. This anion, acting as a tetrabasic acid, binds two ribose molecules forming a tetraester with two pairs of adjacent hydroxyl groups.

In this way boric acid, or more precisely its ion in alkaline solution, stabilizes the desirable product of formose reaction. In addition, by the same mechanism, i.e. by binding to adjacent hydroxyl groups, it stabilizes glyceraldehyde in its enolate form. So both the product (ribose) and the crucial reactant (glyceraldehyde) are stabilized by binding to borate. Could this facilitate the synthesis of ribose, a pentose with an overall yield in ordinary formose reaction of less than one percent?

Every hypothesis must be proved. And what is more, hypothesis of prebiotic synthesis must be proved in realistic environmental conditions. It is not enough to add boric acid to a reaction mixture, because there is no free boric acid in nature (as there are no free fatty acids in fascist Italy, joked Nobel Prize winner Leopold Ruzicka at a scientific congress in that country). Moreover, as calcium hydroxide is also present in the mixture it should react with boric acid as well.

This crucial experiment was performed in 2004 by Alonso Ricardo at the University of Florida as a part of his PhD thesis and published the same year in the journal *Science* under the title "Borate minerals stabilize ribose." For "borate mineral" he chose colemanite, a usually white mineral of vitreous luster formed by crystallization from water during its evaporation (evaporite). But from a purely chemical position, colemanite is a calcium salt of triboric acid, $Ca[B_3O_4(OH)_3] \cdot H_2O$. It is slightly soluble in water, 0.82 g/L, to be precise.

"When the reaction was made in the presence of calcium hydroxide, a simple visual inspection gave some clue of the composition," the young American scientist described results of his experiments on "the synthesis of pentoses under alkaline conditions." "The solution which was originally transparent, turned clear brown after 20 minutes and after one hour dark brown, which, in the case of sugars, is an indicator of decomposition with branching (known as caramelization or browning)." It is apparent that sugars were formed but they didn't persist. But what happened when boric acid was added, in the form of colemanite?

"When boron was introduced as the mineral colemanite into the reaction mixture, visual inspection indicated that the decomposition process was either absent or slow," added the American scientist. "Browning was not observed even under long incubation periods." This means that borate ion stabilized ribose, as well as other pentoses, and protected them from decomposition.

After completing his measurements Alonso was able to report half-lives of pentoses in the reaction mixture. The most stable was ribose, regardless of whether the borate was added or not. If only calcium hydroxide was present, half of the synthesized ribose molecules decomposed after 291 minutes, that is about five hours. But if colemanite was added, its half-life increased to 2700 minutes, or 45 hours or nearly two days. Likewise, ribose was far better stabilized than other six pentoses, in both experiments. Is this enough for synthesis of nucleic acids to occur? Is this the answer to why nucleic acids are composed of ribose, and not of some other simple sugar?

This is the first question. The second one is how ribose started binding to nucleic bases and especially to phosphoric acid.

The phosphorus enigma

Synthesis of organic phosphates, that is esters of phosphoric acid, is quite simple. All you need is pyrophosphoric acid or any of its higher polymers. If you combine it with an alcohol or a sugar (sugars are alcohols!), an ester bond will form immediately by release of one or more phosphate moieties. This is the reaction on which a large majority of all biochemical processes is based. The famous ATP is nothing other than an ester of triphosphoric acid or adenosine triphosphate. Usually it transfers one or two of its phosphoric-acid groups to other molecules, turning itself into adenosine diphosphate (ADP) or adenosine monophosphate (AMP) in the process.

"Phosphate esters and anhydrides dominate the living word," were the opening words of the 1987 *Science* article "Why Nature chose phosphates" by Morris Loeb, Professor Emeritus at Harvard University, who published it under the name of F. W. Westheimer. "The genetic materials of DNA and RNA are phosphodiesters. Most of the coenzymes are esters of phosphoric or pyrophosphoric acid. The principal reservoirs of biochemical energy [adenosine triphophate (ATP), creatine phosphate, and phosphoenolpyruvate] are phosphates." In other words:

Whereto shell I thy worth compare,
Whose actions so admired are?
No substance knowne is like thee
In strength, in virtue and degee.

– to quote John Woodall, a seventeenth-century English poet. (He composed this poem to celebrate the glory of mercury; I took the liberty to change "medicine" into "substance.")

But there is a problem here. Neither free phosphoric acid nor its polymers exist in nature. (Phosphoric acid polymerizes easily enough through heating. A trivial term for diphosphoric acid, pyrophosphoric acid, points to that fact.) They are immediately neutralized by calcium forming insoluble minerals. Apatite and hydroxyapatite are not soluble in water. You cannot obtain phosphoric acid by cooking bones.

In my youth my older colleagues complained against "new chemistry," a chemistry based on instrumental methods and computer modeling rather than direct chemical knowledge, experience and intuition. (It once happened that one of the "new chemists" performed a quite complex X-ray analysis of a yellow powder which was found in his flask following a certain reaction only to discover that it was just plain sulfur. "It is well known that in this reaction sulfur is usually formed," a more experienced colleague chided him.) Nevertheless, experience and intuition is instrumental in solving the problem of formation of the first organic phosphates and even that of the first nucleic acids – and it is no wonder that many hypotheses and theories were developed to cope with this problem.

"We have phosphorylated nucleosides at 160 °C with either phosphoric acid or its monobasic salts," reported Cyril Ponnamperuma in the first of his two consecutive papers published in the May 1968 issue of *Nature*. "Conditions can be created in which inorganic phosphates act as phosphorylating agents," runs the subtitle to both contributions. "Such reaction may have occurred in prebiotic chemical processes." Great!

If apatite, $Ca_5(F,Cl)(PO_4)_3$, and hydroxyapatite $Ca_5(OH)(PO_4)_3$, have no phosphorylating power then other salts of phosphoric acid do. It has been repeatedly said that phosphoric acid, H_3PO_4, is a tribasic acid, which means that it forms three kinds of salts, such as the three "sodium phosphates," Na_3PO_4, Na_2HPO_4 and NaH_2PO_4. The last two salts are actually acids ("acid salts"). And being a kind of phosphoric acid, they could possibly

2 250

ice age (Huronian glaciation)

produce its polymers upon heating. Those polymers are further apt to esterize nucleosides.

And what was the result of these experiments? They showed that pyrophosphoric and tripolyphosphoric acids are not produced exclusively from phosphoric acid. They can also be produced from sodium, ammonium and even calcium hydrogenphosphates by heating them above the boiling point of water. And all of them are capable of producing phosphorylated adenine (nucleoside with purine base) and uridine (nucleoside with pyrimidine base) when those nucleosides are added to the reaction mixture. After two hours of heating at 160 °C with calcium dihydrogenphosphate, $Ca(H_2PO_4)_2$, 10.5% of uridine was converted into its monophosphate. It was also discovered that nucleosides are phosphorylated by action of tripolyphosphates rather than that of diphosphates.

And was the problem of phosphorylation on the primitive Earth finally solved by these experiments? No, it wasn't. The process is quite observable in a test tube but it is difficult to imagine such an event taking place on our planet in its youth. Phosphoric acid, as well as its "acid salts," may only be formed in an acidic environment. There is no indication of acid rain or the like on the primitive Earth. Darwin's soup certainly wasn't sour, but rather we'd find it quite alkaline, if we bear in mind that primordial atmosphere contained a lot of ammonia. And finally, those phosphoric salts had to meet ribose, and ribose, as shown, had formed at alkaline conditions (pH ≈ 9). But where is the solution to this problem?

Back to geology! Primordial Earth's atmosphere was not only a product of gases from the solar nebula which coalesced in the orbit around Earth but it was also a result of volcanic activity. Molten lava, as it cools, produces basalt. And in lava, as well as in magma, phosphorus can be found, just like in fish bones – and phosphorus is good for the origin of life.

There are many kinds of basalts, but these igneous rocks generally have a high content of silicon dioxide (about 50%), followed by aluminum oxide (about 18%) and then oxides of calcium, magnesium, iron, sodium and potassium. (Note that analysis of rocks is frequently given as oxide content.) The content of phosphorus varies. Usually, there are a few tenths of a percent of phosphoric oxide, but in nephelinite, ignous rock named after mineral nepheline, there is 2% P_4O_{10}, due to high abundance of alkali metals. And now, let's look at the experimental evidence.

112

The experiment in question was performed by four Japanese scientists and published in the August 1991 issue of *Nature*. Their first step was to prepare an artificial magma by mixing fine powders of basalts and calcium phosphate. This "raw material" was then placed in a tube running through an electric furnace heated to about 1300 ºC. At this temperature the mixture of rocks and minerals melted entirely. On one end of the tube a steam flow was provided, and on the other end the products of reaction of superheated water vapor and molten rocks were cooled by a water jacket. The condensate was then conserved in an ice bath for further analysis.

At the end of the experiment, phosphorous-31 NMR spectroscopy and anion-exchange column chromatography uncovered many kinds of different phosphoric acids or, more precisely, their ions. Japanese scientists also analyzed condensates of volcanic gas collected from fumaroles. They soon discovered orthophosphate (i.e. ordinary phosphate) and two of its polymers, pyrophosphate and tripolyphosphate. "The vigorous release of gas from the interior of the Earth at a high temperature... would have permitted large-scale production of polyphosphates on the primitive Earth," they concluded.

Afterwards, the phosphoric fumes precipitated as phosphate minerals on the sea bed which then reached magma by action of plate tectonics, to be finally released by volcanic activity into the atmosphere as P_4O_{10}. "Thus, phosphate must have been continuously recycled by volcanic processes throughout the Earth's history," the Japanese scientists stated at the end of their article, demonstrating very clearly that geological fate of phosphates is similar to geological history of carbonates; they are also recycled through plate tectonics and volcanic activity. But phosphates and polyphosphates have to be soluble – and they are not. When a chemist is faced with a "not" he has to think of some catalyst.

This catalyst could be urea or carbamide, the first synthesized "organic substance." It is well known that urea facilitates phosphorylation reactions, and so the first ingredient of *mixtura mirabilis* is defined. The other ingredients are ammonium chloride and ammonium hydrogencarbonate, salts alchemists used to prepare from urine slurry. A mixture of those three components (urea, ammonium chloride and hydrogencarbonate) is capable of mobilizing a phosphate group from hydroxyapatite even and then binding it to uridine. After heating it to the temperature of boiling water, 16% of uridine is phosphorylated.

2 190

ice age (Huronian glaciation)

This experiment, published in 1971, was repeated two years later by the same author (L. E. Orgel) though hydroxyapatite was replaced by a magnesium mineral called struvite, $MgNH_4PO_4 \cdot 6H_2O$. The reason for this was that struvite was presumably formed more easily in primitive Earth conditions than hydroxyapatite. "If ammonia was abundant on the primitive Earth, the evaporation of prebiotic lakes or tide pools could have led to the formation of solid films containing struvite, urea, and other organic compounds including, perhaps, nucleosides and nucleotides," concluded G. J. Handschuh and L. E. Orgel in their February 1973 paper in the *Science* journal.

The key word in the quoted sentence is "perhaps." Perhaps our bone soup was not made from bones, or hydroxyapatite, but from its magnesium counterpart, struvite. However, it isn't only phosphorus that is good for your brain; magnesium can serve as well.

RNA world

In 1970s everything seemed clear and simple, *Gloire à Dieu, honneur au Roi, salut aux Armes*, as they used to say in France during the reign of Louis XIV. There were three realms of biochemistry (and many small principalities and duchies): DNA, RNA, and proteins. DNA was the database, RNA its interpreter, and proteins were the executive staff. Or, in another analogy, DNA was the constitution, RNA the state administration and proteins the industry. To be brief, and much closer to chemistry:

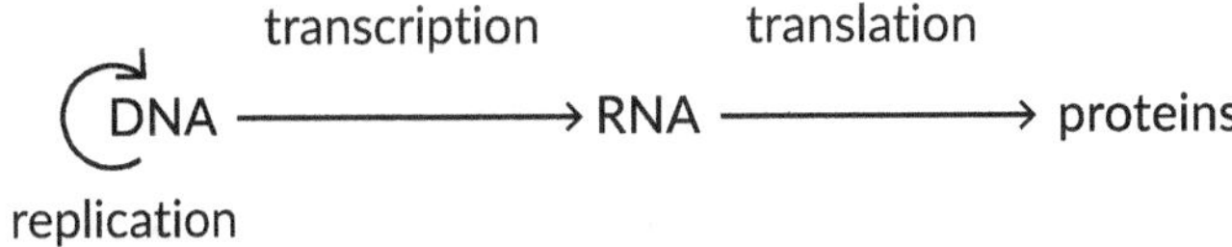

This simple scheme, known as Central Dogma, clearly shows that genetic information is transferred by a copying mechanism from DNA to RNA and further on to proteins, which are mostly enzymes. Enzymes are catalysts responsible for driving all chemical changes in a cell, including the processes of replication, transcription and translation. DNA cannot function without proteins, and proteins should never be synthesized without the aid of nucleic acids and *vice versa*:

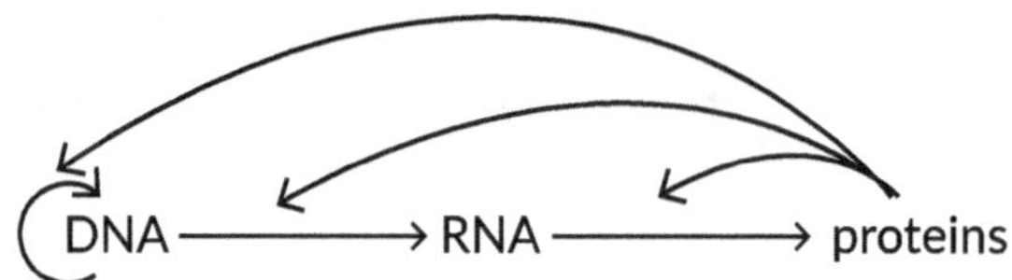

This is the essence of Central Dogma. "Most importantly, both these latter arrows are unidirectional, that is, RNA sequences are never copied on protein templates; likewise, RNA never acts as a template for DNA," James D. Watson wrote in his 1970 textbook *Molecular Biology of the Gene*. But never say never!

RNA does act as a template for DNA in case of some viruses, including the infamous HIV. They are called RNA-viruses because their genome is not composed of DNA but of RNA. In HIV viruses' capsid, in addition to the RNA molecule – its "chromosome," there are two molecules of reverse transcriptase, an enzyme which reverses the first "arrow," DNA → RNA into DNA ← RNA, to further incorporate newly synthesized DNA into a cell genome. But that is not all. Some RNAs may act as catalysts (RNA enzymes or ribozymes). The mechanism of life is much more complex than anyone dared to imagine.

This revelation arrived in 1982, when Thomas R. Cech, professor at University of Colorado, made a discovery that "a shortened form of the self-splicing ribosomal RNA (rRNA) intervening sequence of *Tetrahymena thermophila* acts as an enzyme in vitro," as summed up in the January 1986 issue of *Science*. Or in plain English, the part of RNA molecule found in ribosomes of the named protozoon could catalyze some reactions in a test tube. This part of the molecule is "self-splicing," and is called intervening sequence of RNA, IVS RNA in short.

IVS RNA is 413 nucleotides long. Astonishingly enough, it splits itself without the aid of any enzyme (protein). In doing so it produces two polymers, linear 15-mer RNA and circular 339-mer RNA, -15 IVS RNA. But if a base is added to the reaction mixture in order to increase its pH value to 9, the circular molecule, -15 IVS RNA turns into a linear one, to be again cyclized by releasing a tertramer. This new circular RNA, now 395 units long (-19 IVS RNA) may be opened again to produce a linear molecule of L-19 IVS RNA of the same length.

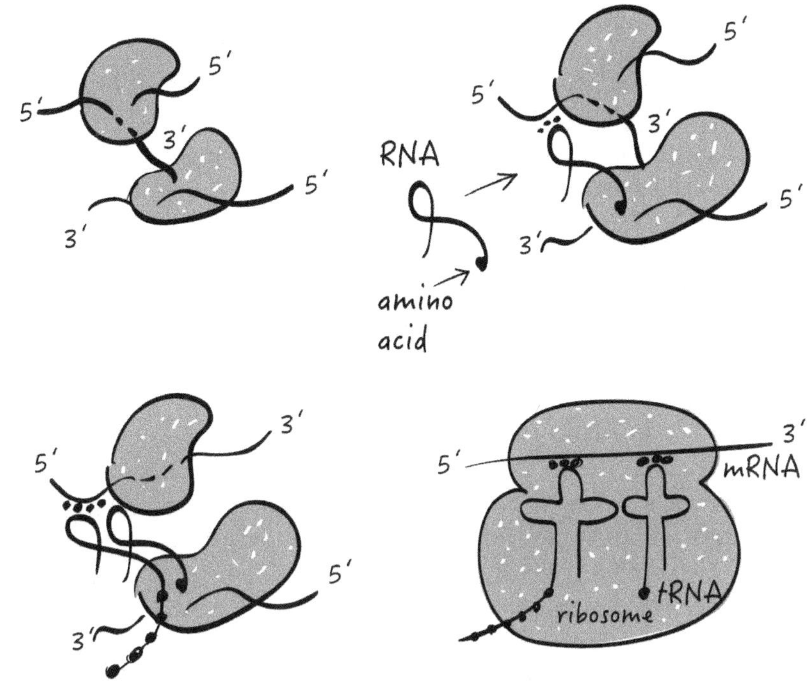

Evolution of RNA world into modern central-dogma world; self-replication of RNA (RNA→RNA) gradually evolved into protein synthesis on a ribosome and mRNA template (RNA→protein). Note that first RNAs were both template and catalysts. (Adapted after W. R. Taylor, *Phil. Trans. Roy. Soc. B* **361** (2006) 1751–1760.)

"We have now tested this [our hypothesis] by adding oligonucleotide substrates to the L-19 IVS RNA," wrote the American scientist. "We find that each IVS RNA molecule can catalyze the cleavage and rejoining of many oligonucleotides." In other words, L-19 IVS RNA is capable of destroying ("cleavage") and synthesizing ("rejoining") RNA molecules, and "Thus, the L–19 IVS RNA is a true enzyme." It acts as an RNA-polymerase – even though it is not a protein. (Each enzyme catalyzes a specific reaction, the forward as well as the reverse one. This is so because enzymes, as other catalysts do, merely establish a chemical equilibrium.)

This discovery and a similar discovery by Sidney Altman of ribonuclease-P, a ribozyme engaged in the synthesis of tRNA, fascinated and motivated F. H. Westheimer to write the article "Polyribonucleic acids as enzymes," which was published a month after Cech's paper in the *Nature* journal. In that article he did not report any new experiment or give any new interpretation of the published results. He simply drew some new conclusions. If RNA is capable of synthesizing and being cleaved by itself, then proteins are not necessary. Life could be functional without them as well. "Bass and Cech have wondered whether the ribozymes may not have preceded protein enzymes," wrote Westheimer. "As ribonucleotides [RNA] apparently possess both the properties of self-replication and those of catalysis, could these molecules have allowed a primitive living system to function without either DNA or protein?" The very first world was an RNA world.* And RNA ruled alone.

The seven arguments

TThere are many arguments in favor of this hypothesis, or – to be scientifically precise – there are exactly seven arguments in favor of an RNA world, as presented in 1989 by Gerard F. Joyce. The first argument is that the mechanism of self-replication of RNA is much simpler than the pathway of DNA replication. The second is the above mentioned discovery of ribozymes. The third argument is based on the wide diversity of RNA functions, suggesting its ancient origin.

The two following arguments point to the remnants of ancient ribozymes in contemporary proteins (enzymes). Coenzymes, nonprotein yet active parts of enzymes, are mostly nucleotides or molecules resembling them (the fourth argument). One such compound, amino acid histidine, which is also frequently found in the active site of enzymes, is synthesized from phosphoribosyl pyrophosphate and adenosine triphosphate (ATP), quite similar to the synthesis of RNA. Does this argument, the fifth one, point to the fact that histidine is a remnant of an ancient ribozyme?

The sixth argument states that RNA is phylogenetically older than DNA, because deoxyribonucleotides, DNA "bricks," are synthesized by reduction of ribonucleotides, building blocks of RNA. And what about the seventh, the last argument? Joyce pointed to the fact that deoxythymidylic acid is formed by methylation of deoxyuridylic acid (pU → pT).

*Despite of launching the RNA world theory in 1986 (Westheimer, Gilbert), it can be traced back to 1962 when Alexander Rich hypothesized that RNA might have played both catalytic and genetic roles in early forms of life.

The theory of RNA world was later strengthened by some new striking evidence. Ribozymes are not enzymes which deal with RNAs only, but they do so with proteins as well. Peptidyl transferase, the key enzyme in the synthesis of proteins on ribosomes, is actually ribosomal RNA! Its function is to make new peptide bonds between amino acids bonded on one tRNA molecule (aa-tRNA) and the growing peptide chain attached to the other (peptidyl-tRNA). "The striking similarities between the sequences containing the key catalytic elements found in peptidyl-transferase active site of the ribosome and the sequences of in vitro-selected RNAs [their model compounds] having related activities make it clear that the appearance of a small RNA domain capable of catalyzing [as] peptidyl transferase was a plausible first step in the evolution of protein synthesis on the ribosome," wrote the five American authors of the paper "The structural basis of ribosome activity in peptide bond synthesis," published in August 2000 issue of *Science*.

This means that the first living things were clumps of RNA molecules possessing catalytic properties from which ribosomes latter evolved. At first, this new peptidyl-transferase activity produced peptides of random sequence, but these protopeptides eventually found their role in the evolution of first life. "Such peptides could have enhanced the stability of protoribosome and other early ribozymes as the more sophisticated peptides of the present-day ribosome appear to do" reads the paper's final conclusion.

To sum it up, RNA is older than both DNA and proteins, and what is more, it may integrate functions of both classes of biomolecules, DNA and proteins. Life did not begin with catalytic proteins, as Oparin believed, but with self-replicating ribonucleic acids, cell components of virtually no interest to the Russian scientist.

But there are two flaws to this argument. RNA is not chemically stable; it is – in contrast to proteins – easily cleaved. The second problem is its prebiotic synthesis. This synthesis of ribose by a borate catalyst, and its phosphorylation by means of urea is just one side of the story. The other one is much simpler; it's clay.

A clay pond

As a boy I was very fond of a book written by German chemist Hermann Römp titled *Chemistry of the Future*. What are we going to do when ore deposits get depleted? For example, numerous rich deposits of bauxite exist today, but everything comes to an end eventually. And then... we will switch to clay. Cook clay with concentrated hydrochloric acid and you get aluminum chloride. If you treat this salt with water steam, hydrochloric acid will regenerate and alumina (aluminum oxide) will be produced as raw material for the production of aluminum in an electric furnace. In the nineteenth century there lived a man who was paid by the government five Forints a word to propose a new Croatian scientific terminology. An idea occurred to him, probably inspired by desire to increase his pay, of introducing "domestic" words for chemical elements. Accordingly, phosphorus was called *kostik* (from *kost*, bone), and aluminum *glinik* (from *glina*, clay). Yes, that's accurate: clay is very rich in aluminum.

Clay, clays or more accurately clay minerals are produced by weathering, hydrothermal processes and the like; in other words, by action of water on silicate rocks. The most common such mineral is kaolinite, the chief constituent of clays used for china and pottery. However from the perspective of this book the most interesting clay mineral is montmorillonite, named after Montmorillon in the French Alps. It is the chief member of montmorillonite group, also called fuller's earth or smectite group.

Writing its chemical formula down doesn't make much sense. It might be better to say that it is composed of silicon, oxygen, aluminum, magnesium and iron as well as of hydroxyl groups and water molecules. Its structure may be described as being composed of two kinds of layers. The first layer is made of octahedral aluminates; that is aluminum ions with two hydroxyl groups attached and interwoven with magnesium, iron and oxygen ions. This layer is sandwiched between two layers of polymerized tetrahedral silicates, forming a negatively charged platelet. These negatively charged platelets are stacked one on top of another, like a deck of cards. But why don't they repeal each other? It doesn't happen because of a positive charge dispersed in the form of cations dissolved in layers of water which separate the platelets.

Montmorillonite is formed by weathering of volcanic ash, and so it could have easily formed on the primitive Earth. It was even found on Mars, and this is an argument which suggests that life might have once originated on that planet. But first we have to find out how life originated on *this* planet of ours.

There are few cat owners who are aware that "sand" (bentonite) in their pet's toilet is actually montmorillonite, and they certainly don't know why

2 023

2 010

5-10 km impactor
(Vredefort, South Africa)

this "sand" is so suitable for disposing cat litter. It is a very good adsorbent as well as a powerful catalyst. Its catalytic properties may be further enhanced if it is rinsed with an acid or exposed to salt solutions. It is because cations in its water layers which separate alumosilicate platelets are easily exchangeable. In doing so it is even possible to prepare montmorillonite capable of linking nucleotides into molecules of ribonucleic acid.

In other words, clay mineral acts as a catalyst, like an enzyme (or ribozyme) RNA polymerase. If Na^+-montmorillonite is mixed with RNA monomers activated with imidazole, Im (ImpA, ImpG, ImpU, and ImpC) in a slightly basic solution, after a few days RNAs of various lengths will appear. But that is not all!

I already mentioned the enormous number of possible protein molecules. If you have this many amino acids that are to be arranged in this long long peptide chain, the number of combinations is... A similar calculation can be made for nucleic acids. The 50-mer RNA molecule, one consisting of 50 nucleotides, may be synthesized into 10^{30} different versions! If we were to synthesize just one molecule per possible version, we'd end up with a mass of 10^{10} grams – or 10,000 tons! It is absolutely impossible that any natural RNA could have been synthesized by a unidirectional, random process.

But then again, synthesis of RNA on montmorillonite is not a random process. Or it may more precise to say that its synthesis is not purely random. For a reaction between activated adenosine and cytidine (ImpA and ImpC) eight dimers should be predicted (AA, CC, AC and CA – but the nucleotides could be bonded in two ways with respect to ribose moiety). Consequently, there are 32 trimers, 128 tetramers and 512 pentamers. But this combination game is not reflected in experimental results. Surprisingly, only 10 (instead of 32) trimers were observed in the reaction mixture. Similarly, there were five tetramers and only four pentamers. Contrary to the same yield expected in random synthesis, clay-catalyzed synthesis is highly selective even in this respect. For instance, yields of the five tetramers varied in a rather wide range from 5.5 to 34%.

But the riddle still remains unsolved. Where did the activated nucleotides (ImpA, ImpC, etc.) come from? Something had to be there before them.

Did life begin in hot volcanic water?

1 950

Mill. yrs from present

6. THE GRILL

What do broiling of meat and the origin of life have in common? They are both based on hydrothermal processes – if we accept the iron-sulfur-world theory and a strange notion that the very first life had nothing to do with proteins and nucleic acids.

There are no people on Earth who are unfamiliar with the process of grilling. And why is that? Even the *Homo sapiens neanderthalensis* from Krapina knew how to use fire – this indicates that he used to prepare meat, and that he did it in the simplest way possible, by exposing it to glowing charcoal. Modern soldiers who find themselves in dire circumstances commonly prepare their food in this way. During their retreat from Moscow, soldiers of Napoleon's *Grand Armee* broiled meat of dead horses over open fires. As there was no salt to be had, and much less other spices used in the refined French cuisine, they used gunpowder instead. There is saltpeter (potassium nitrate) in it, and saltpeter is salty. (Did they salt their meat before or after they broiled it?)

This kind of cuisine was well known to partisan fighters hiding in the forests of Yugoslavia during World War II. And they also boiled a broth of beech leaves to go along with whatever else they could find.

- Why did our partisan resistance fighters eat beech leaves? – The professor asked his student.

- Because they had nothing else to eat – she answered, quite naively.

- And ...?

- Because they were very hungry.

- Didn't it occur to you that beech leaves are rich in cellulose fibers, and fibers stimulate intestine peristaltic?

- If you heard a machine gun burst over your head, you wouldn't have any problems with your peristaltic.

But there is another way of preparing meat over open fire. A piece of meat may be skewered on one end of a stick. During my stay in the United States, I used to hike with my colleagues, mostly Poles, to a nearby mountain. Each hiker placed his special Polish brand sausage on the end of a stick and roasted it over open fire. In my family there is a similar traditional custom. We put a whole lamb on a spit.

Spit-roasted lamb is a popular Croatian delicacy, especially in its southern regions (in the North we do the same with pig sucklings). During the Communist era, it was a customary delicacy for high party and government officials, which led to a nickname "lamb heads," possibly because they preferred that portion of lamb's body (which is actually the tastiest portion, even though not many know this). Jiří was a Czech language translator of my father's books and his close friend. He spoke Croatian so fluently that I never thought of him as a Czech unless he made typical Czech jokes. These jokes – also popular in other countries of central Europe – are not simple puns or gags, but actually complete humorous yarns about the other, less "official" side of life. (During the Russian occupation in the 1970s Jiří sent us a Christmas card showing Russian tanks running through a snowy pine forest, quite romantic.) To impress our guest, we prepared our favorite, lamb on a spit at our old mountain family house. Our family and friends were all gathered in the yard, hoping to snatch the choicest bits of the broiled lamb, but Jiří shut himself into the house and did not venture outside. He claimed to suffer from stomach problems.

Many months later, after he returned to Prague, he finally made a confession to my father. His stomach was quite all right, but he was made ill by the sight of a little cute lamb impaled on a horrible big stick. (In our civilized world, we desperately strive to prepare our meat as discretely as possible, out of vain hope that we'll forget how it ultimately came to our table.)

Meat heating or, to be more precise, the broiling of meat over an open fire, has something in common with another crude and a very ancient process. The process of broiling is based on heating the meat with hot air (enriched with smoky gases and particles) and thermal, infrared radiation emanating from the glowing coal. This heat causes body liquids, water and melted lard, to circulate within the body and it is this circulation which broils the meat. (It is interesting how spitted lamb is salted. Salt is placed in just a few places inside the body before broiling, as it will disperse itself uniformly following three to four hours of such circulation.)

1 900

1 890

formation of
super-continent
Columbia (Nuna)

This closely resembles processes present in the Earth's crust. In addition to this, it is also constantly baked by the heat of molten rocks in the mantle. As in the body of a lamb, there is constant circulation of hot water below our feet and this water occasionally finds its way to the Earth's surface, creating hydrothermal vents. Not all of them are as spectacular as geysers in the Yellowstone national park and they most commonly occur in the form of hot springs where they are commonly used for spas. Correspondingly, there are many hydrothermal vents in the deep, at the bottom of the ocean, in the abyss… where life began. Perhaps.

Where nobody had expected it…

In 1844, British naturalist Edward Forbes, three years after his expedition to the Aegean Sea on HMS *Beacon*, quite reasonably affirmed that there was no life in the ocean at depths lower than 300 fathoms (550 meters), simply because sunlight could not reach waters below that depth. As all ocean life is dependent on phytoplankton, and phytoplankton are green plants dependent on sunlight, there is no food available for any organism living below the boundary of 300 fathoms:

As we descend deeper and deeper into this region, the inhabitants become more and more modified, and fewer and fewer, indicating our approach to an abyss where life is either extinguished, or exhibits but a few sparks to mark its lingering presence.

But what may sound reasonable does not necessarily have to be true. There are many "unreasonable" miracles in the sea, and in my country the expression "a sea miracle" (*morsko čudo*) means something utterly improbable yet still true.

"The sea has many voices,/ Many gods and many voices," wrote T. S. Eliot, and these voices are "Often together heard." In 1818 – twelve years before Forbes put forward his azoic hypothesis – Sir John Ross, Admiral of the Royal Navy, undertook his first expedition to Arctic seas where, at the depth of 1000 fathoms (about 1800 meters) he discovered a multitude of creatures living in the mud at the bottom and thereby concluded, "thus proving that there was animal life in the bed of the ocean notwithstanding the darkness, stillness, silence, and immense pressure produced by more than a mile of superincumbent water." In 1957, more than a hundred years following this momentous discovery, a Swiss physicist named Auguste Piccard completed building his first *bathyscaphe*, "ship of the deep," and three years later his latest vessel *Trieste* reached the ultimate ocean floor of the Earth, the Marianas Trench in the Pacific, located southwest of Guam in the Mariana Islands. And what did he find at the depth of 11 kilometers? A flatfish.

It was really unbelievable. In 1856 David Page, a geologist, wrote: "According to experiment, water at the depth of 1000 feet is compressed 1/340th of its own bulk; and at this rate of compression we know that at great depths animal and vegetable life as known to us cannot possibly exist – the extreme depressions of seas being thus, like the extreme elevations of the land, barren and lifeless solitudes." But this is not really true. We should expect to find life everywhere, and there is no such thing as a sterile place on Earth.

Am I assuming too much? Sterilizing conditions certainly do exist on our planet. Is it not a proven scientific truth that no germ can survive prolonged immersion in water at above its boiling point? This is a fact, discovered by Louis Pasteur in the nineteenth century, which forms the foundation of our trust in healthiness and safety of canned food. To can food means to have it "pasteurized" which is shorthand for sterilization by heating (usually at 60 – 90 °C). There is no germ in existence which can survive such prolonged heating; the upper limit is set at 60 – 80 °C, and that only for a special kind of microbes known as "thermophiles" (temperature-loving ones). But never say never; especially in science.

In my youth I was fascinated by Piccard's bathyscaphe, a submarine vehicle resembling a balloon; there was a hollow sphere made of hardened steel, a console which housed scientists hanging from a gasoline tank whose function was to provide buoyancy. In the remarkable "international geophysical year," 1957/8, many scientific miracles were performed, such as the first "artificial Earth satellite," the Russian *Sputnik*, which was launched into the low Earth orbit and initiated a world-wide public frenzy for new scientific discoveries. Nothing could escape the notice of laymen, and common people were as interested in new scientific discoveries as we are interested in the results of football matches today. But times have changed. Twenty years later, in 1977 nobody noticed a discovery no less unexpected than the discovery of flatfish in Marianas Trench.

Times have changed, technology has moved on and *Alvin*, a new extreme-depth submarine appeared on the scene. Its hollow three-meter

1 830

oxygen had risen
to 10% of present
atmospheric level

titanium sphere housed a crew of three – one pilot and two scientists. The submarine had its maiden voyage in 1968 on a mission to find a hydrogen bomb accidentally dropped from a B-52 bomber near the coast of Spain. *Alvin's* pilots found and recovered the bomb from the sea floor at the depth of 910 meters. In 1977 *Alvin* got itself another job – helping scientists explore underwater volcanoes, or to be more precise, their hydrothermal vents.

Volcanic eruptions taking place on land are quite conspicuous as they tend to produce quite a lot of noise, smoke, ash, and molten lava. Beneath the surface of the sea it's a different story altogether. There isn't much that can be observed from the surface except some hot water – deep under the sea even steam formation becomes very difficult. (This is a natural consequence of high pressure; water boils at 100 ºC at 1 bar, at 200 ºC at 15 bar, that is at the depth of 140 meters, and so on.) In short, water around undersea volcanoes is so hot (300 – 400 ºC) that nothing in it could possibly survive – all fish are immediately converted into fish soup!

This was the expert opinion of the two American scientists, John B. Corliss and John M. Edmond, from Oregon State University and MIT, respectively. In 1977 they used their *Alvin* vehicle to visit hydrothermal vents near Galápagos Islands at the depth of 2000 meters. But instead of a sea floor resembling surface of the Moon, they discovered "The Rose Garden" – a fantastic undersea habitat, vibrant with life: reefs of mussels, crabs, giant clams, anemones, and fish (none of them cooked in the least). And an even more surprising discovery was made later as it was observable only through a microscope.

The discovery was this: a multitude of quite improbable microbes make their home around hydrothermal vents, in structures known as "chimneys" or "black smokers." In 2003 microbiologists from the University of Massachusetts isolated a bacterium from a hydrothermal vent off Puget Sound which kept thriving in boiling water, at 100 ºC exactly (thermophiles have optimal growth at 55 – 65 ºC). This bacterium managed to reproduce at as much as 121 ºC and was able to survive the temperature of 130 ºC for whole two hours. Another microbe, *Pyrodictium occultum* grows fastest at 105 ºC, and another, named *Pyrolobus fumarii* thrives at temperatures of up to 113 ºC. Their enzymes are active at even higher temperatures; they have their optimum at 142 ºC!

Many such highly heat-resistant microbes were found since. In one 2006 paper we are able to read: "At present, about 90 species of hyperthermophilic archaea and bacteria are known, which had been isolated from different terrestrial and marine thermal areas in the world." Moreover, in the same paper Karl O. Stetter from the University of Regensburg stated that he collected more than 1500 strains of hyperthermophilic organisms. They are grouped in as many as 90 species, 34 genera and 10 orders!

Even more important than diversity expressed by 90 species, 34 genera and 10 orders, is that hyperthermophilic microbes are archaea as well as bacteria. This distinction is of immense importance for our discussion, because archaea is a phylum of the oldest organisms on our planet. In contrast to the old division of all living things into plants and animals (the two traditional "Kingdoms"), nowadays there is a more advanced division, based on ribosomal RNA which divides all living beings into "Domains" of Archaea, Bacteria and Eukarya (of course, both Archaea and Bacteria are prokaryotes). This means that archaea from genera *Pyrolobus, Pyrococcus, Thermococcus, Ignicoccus* – as their imaginative names go – are possibly the oldest organisms on our planet. Phylogenetically speaking, the oldest among them is *Nanoarchaeum equitans*, an organism with a cell diameter of only 0.4 μm (the usual size of bacteria is 0.5 – 1 μm) and a genome with less than 500,000 DNA base pairs (the *E. coli* genome has 4.6 million base pairs). It now seems that life didn't originate at the surface of the ocean or in "a warm little pond" somewhere in its vicinity, but at its very bottom, near the hydrothermal vents of submarine volcanoes, as suggested by the discoverers of The Rose Garden. But this is also the essence of the iron-sulfur-world theory launched by the German scientist Günter Wächtershäuser in 1988. His theory is strongly supported by the curious metabolism of these organisms.

Thriving on poison

Bacillus bulgaricus converts milk sugar (lactose) into lactic acid and in doing so produces yogurt. Its relative *Lactobacillus casei* does the same when producing cheese. They both use milk sugar as a source of energy, and lactic acid which is produced alongside it is nothing but metabolic waste, a substance they cannot convert further to obtain energy. Glucose is another source of energy for microorganisms, but it is usually converted into ethanol, common alcohol, thus producing many kinds of beverages. There is also a third kind of microbes (archaea), those who produce methane, the simplest of hydrocarbon gases – a process used in the production of biogas from stable manure and many other kinds of organic waste. (Methane is also produced by intestine microbes, though those of the "bad" kind.

This gas was discovered in the sixteenth century by van Helmont who described the burning of inflate gases by comparing their flame with a rainbow.) This is what is commonly known about "bacterial" products and "bacterial" metabolism. But this is only the tip of an iceberg. "Morphologically, bacteria are simple organisms and are often regarded as a primitive type of cell," should be read in a textbook on microbiology. "However, these organisms are not simple biochemically."

But they cannot be assumed to be simple biochemically just because they morphology happens to be simple, "primitive." In higher organisms cells are specialized, i.e. dependent on each other, but a single cell bacteria or archaea must do everything by itself: "The simplicity and variability of the nutritive requirements of many bacteria also suggest a physiologically primitive cell."

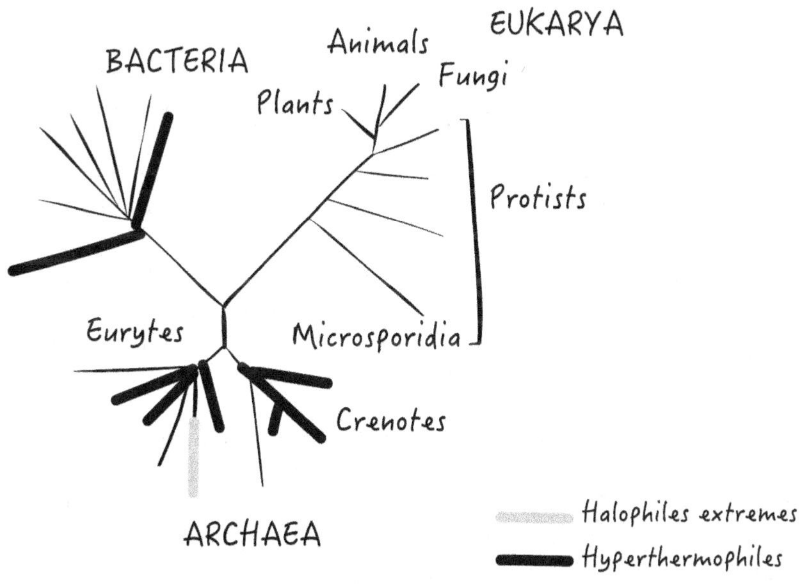

Phylogenetic tree tracing all beings living on this planet, constructed according to their small subunit ribosomal RNA (C. R. Woese et al., *Proc. Natl. Acad. Sci. USA* **87** (1990) 4576–4579). Distances roughly correspond to differences in rRNA sequence. Note that the first organisms were hyperthermophiles.

In other words, there are bacteria and archaea which thrive on complex organic substances, sugars, starch, cellulose and proteins by converting them into simple compounds with one to four carbon atoms, like methane, ethanol, formic, acetic, lactic, and butyric acid. This process is usually called fermentation and microbes which use it are known as heterotrophs.

But there is yet another kind of microbes. They obtain energy by chemical reactions of simple inorganic compounds, using a process known as respiration. These are the chemoautotrophic organisms, and the reactions necessary for their survival are redox reactions. Participation of elemental oxygen (O_2) in respiration is common, but it is not an absolute requirement; there are microbes which obtain their energy by reducing carbon dioxide into methane ($4H_2 + CO_2 \rightarrow CH_4 + 2H_2O$) or oxidizing sulfur into sulfates ($5S + 6KNO_3 + 2H_2O \rightarrow K_2SO_4 + 4KHSO_4 + 3N_2$). Diversity of microbial metabolism is the key to their differentiation, which is also reflected in their names, *Thiobacillus thiooxidans*, *Thiobacillus denitrificans* and the like. But let us return to the deep; let's dive into the hot waters of "black smokers" of Galápagos. What is the natural diet of bacteria and archaea dwelling in The Rose Garden?

Following *Alvin's* return to its mother ship *Knorr*, scientists on board analyzed water samples collected near the hydrothermal vents. In them they found high concentrations of hydrogen sulfide, a gas with very unpleasant smell (similar to that of addled eggs) and as poisonous as hydrogen cyanide. It was obviously produced by a reaction of iron sulfide dissolved in molten rocks with ocean water. This is a well-known reaction which I experienced once when, to my dismay, I poured water on some hot coal ash – a clear proof that coal contains sulfur in the form of sulfides. But what about those microbes from the deep?

As was expected, those microbes were thriving on hydrogen sulfide dissolved in the sea water. They used energy of its conversion to create sulfuric acid ($H_2S + 2O_2 \rightarrow H_2SO_4$), a process with serious consequences for our drainage systems (in which similar microbes can also be found) because plumbing in general is poorly resistant to accumulator acid. But there are even more serious consequences in a quite different field of human endeavor. The discovery of sulfide-oxidizers in ocean depths created new insights into the origin of life on our planet.

Let us now recall Oparin's proposition that primordial Earth was abundant in various complex organic compounds, which were food for his coa-

1 710

widespread appearance
of cyanobacteria

cervate droplets.* This means that the first organisms were heterotrophic, and it took them hundreds of millions of years to develop a process for utilizing solar radiation, having gradually evolved into the first "green" organisms, cyanobacteria. If we follow this line of reasoning, it was necessary for the origin of life to propose the existence of high-molecular compounds, amino acids, sugars, proteins and the like in the primordial ocean. But their existence on early Earth is all but certain, in contrast to low-molecular compounds, such as carbon monoxide and dioxide, ammonia, nitrogen, hydrogen sulfide, methane and phosphoric acid. "The failures of the theories of an origin of life by polycondensation in a 'prebiotic broth' posed the challenge whether it would not be possible to develop a theory of an origin at the level of low-molecular organic compounds," wrote Günter Wächtershäuser in his 2006 paper. "As a solution to this problem, a theory of chemoautotrophic origin from inorganic starting material was developed." This is the already mentioned "iron-sulfur-world" theory (as Wächtershäuser's theory has been dubbed) which, despite starting with simpler substances, is a much more complex one than Oparin's and similar theories. This theory does not propose merely a creation of chemoautotrophic organisms instead of heterotrophic ones in the primordial broth, but the emergence of life from structures having nothing in common with anything resembling a living cell. In the beginning, says Wächtershäuser's theory, there was iron and sulfur and they joined together in iron sulfide...

The slurry of life

As a university student I developed a taste for white wines, which I previously regarded with suspicion as quite strange sour liquids, particularly when compared to strong hearty red wines I used to drink from a traditional wooden barrel owned by my grandma, who used to make wine from dark grapes grown in sunny Dalmatia (to me wine was never a "drink," I always considered it a normal part of any healthy diet). But after just two glasses of white wine in restaurant Plješivica (named after a central Croatian mountain topped with vineyards), I came home and was terribly sick throughout the night. "It is pure poison," I said to a guest at nearby table the following

---

*Oparin also considered chemoautotrophs, but only as a transition phase from heteroautotrophs to photoautotrophs. According to the Russian scientist, this phase did not last very long because of the shortage of inorganic compounds.

evening. "Don't drink it!" And he didn't. He lost all desire to do so after he poured some mineral water into the wine to prepare the popular *gemišt*: the liquid changed its color and grey slurry formed immediately.

My chemist's brain reacted immediately. I often observed similar slurries during my laboratory training; they usually formed through reaction of hydrogen sulfide or soluble sulfides with heavy metals, copper, iron and – God forbid finding it in wine! – with lead.* Formation of lead sulfide precipitate was used by German chemists after the Great War to separate gold from sea water in a futile attempt to convert the world's ocean into a gold mine providing funds for war reparations imposed by the Treaty of Versailles. Reaction of ions of another heavy metal, iron, with soluble sulfides had no influence on world politics however, but it did exert great influence on biochemistry.

Reaction of iron(II) ions with inorganic sulfides is quite simple ($Fe^{2+} + S^{2-} \rightarrow FeS$), but what happens when Fe^{2+} reacts with organic sulfides, compounds with general formula R–SH?

The answer is simpler still, because their chemistry is simple as well. Sulfur likes to bind with iron irrespective of what is otherwise bound to it, such as another hydrogen atom, as in the molecule H_2S, or a methyl group, as in the molecule CH_3SH – a bad-smelling substance which is commonly added to natural gas in order to signal leakage. In a series of simple experiments, not much more complicated than the experiment of preparing *gemišt* from this "great" wine of my youth, chemists made dozens of combinations of Fe^{2+}, HS^-, S^{2-} and R–S$^-$ ions (here R stands for any group of atoms) producing substances with some very unusual structures and properties.

By combining these ions clusters are formed. They are not clusters of grapes used in the making of wine mentioned at the beginning of this section, but clusters of atoms. There are many clusters composed of gold, silver and copper atoms, but the clusters we're referring to are those consisting of iron and sulfur atoms, all possessing a mandatory organic component. Their typical structure is a cube; four of its vertices hold iron atoms, and the

*Note a very subtle equilibrium among H^+, S^{2-} and the ions of heavy meals, e.g. Fe^{2+}. At very high H^+ concentration (low pH) hydrogen sufide, H_2S, is evolved. At pH higher than 6.5 various soluble complexes ($[FeSH]^+$, $[Fe_2(HS)]^{3+}$, $[Fe_3(HS)]^{5+}$) of iron are formed, and in basic solutions (pH >7) Fe^{2+} combines with free S^{2-} making insoluble FeS. It is thus quite possible that both Fe^{2+} and S^{2-} were already present in this wine, but they did not precipitate because wine was too sour (acidic). By adding carbonated mineral water (*i.e.* bicarbonate buffer, HCO_3^-/H_2CO_3), pH rises forming FeS particles.

other four hold atoms of sulfur, also bonded to R. Their general formula is $Fe_4(SR)_4$, and they exhibit some very interesting properties.

First of all, they easily accept and release electrons (note that crystals of iron(II) sulfide are semiconductors!). This is easy to account for, because iron exists in two oxidative states, iron(II) and iron(III). These states are easily exchangeable by releasing or accepting an electron ($Fe^{3+} + e^- \leftrightarrows Fe^{2+}$). As there are four iron atoms in a cluster, there are five combinations of Fe(II) and Fe(III) states, namely $Fe(II)_4$, $Fe(II)_3Fe(III)$, $Fe(II)_2Fe(III)_2$, $Fe(II)Fe(III)_3$, and $Fe(III)_4$. One oxidation state is easily transformed into another by electron transfer, and as electron transfer is just an alternative name for redox reaction, iron-sulfur clusters are considered to be redox catalysts.

A multitude of iron-sulfur clusters has been prepared in laboratories already, and many reactions have been studied by using their power of catalysis. They catalyze oxidations of thiols (RSH) into dithioethers (RSSR), interchanges of R groups (e.g. RSH into R'SH) or ester synthesis (esterifications). But what is most interesting is that such reactions do not occur in test tubes only, but in living cells as well.

The story of iron-sulfur clusters began in the 1970s with the discovery of iron-sulfur proteins. Something quite inorganic was found at the core of those proteins – clusters composed of iron and sulfide ions bonded via S-bond to side chains of the protein (or to be more specific, to the side chain of its cysteine residues). The simplest one is [2Fe:2S] cluster and the most common is the one just described, a cube with four Fe and four S vertices, [4Fe:4S]. These proteins are involved as catalysts (enzymes) in photosynthesis, cell respiration, nitrogen fixation, and in many other processes. They are hydrogenases (i.e. they reduce H^+ into H_2), nitrogenases (they convert N_2 into NH^+_4), and also enzymes involved in oxidation of sulfur. They are the key enzymes found in bacteria *Thiobacillus thiooxidans* and *T. ferrooxidans*, which are used for "bacterial leaching" of copper sulfide ores, a process which converts them into soluble copper sulfate.

Wächtershäuser's iron-sulfur-world theory stems quite naturally from this. There used to be quite a lot of dissolved iron in the primitive ocean and – this is important to note – this iron was dissolved in the form of Fe^{2+} ions. There was no free oxygen in the atmosphere yet, and thus Fe^{2+} couldn't be oxidized into Fe^{3+}, a form of iron which easily combines with oxygen forming insoluble precipitates. (When oxygen eventually appeared, ocean iron precipitated – forming large fossil deposits.) Though this primitive ocean was free of life at the time, it wasn't free of organic substances and hydrogen sulfide which developed – even more actively than today – from the multitudes of volcanic fissures. As a result of these processes iron-sulfur clusters were formed – the first catalytic particles. From these particles the first living cell eventually evolved.

This is a very convincing hypothesis; what is life but a catalytic activity of enzymes in the cell? But the story is not as simple at all – as always, the devil hides in the details.

Life driven by the formation of pyrite

Cooking is a simple process, but nobody knows exactly what is happening in the pot. Anyone with some education in chemistry would say that cooking is a process of denaturation of proteins, dissolving of mineral salts, melting of lard, and emulgating it with the aid of albumins or lecithin, although which of those substances combine with what exactly, and moreover by which chemical equations we should describe these reactions in the pan or pot, nobody can say with certainty. The reason for this is not because it is difficult to propose reactions in the pot but quite the opposite, because they may be proposed too easily. There are many possibilities, but which is the right one? It is a valid scientific question.

And so it is with the origin of life. There are many possible reactions with FeS particles, but which are the ones that led to the emergence of life, to the first living cell? And of even greater importance is the question of the key chemical reaction which produced the energy for this process. This last question is in the scope of chemoautrophic theory, for it presumes a process which draws energy from the simplest reactions.

I already mentioned that the key reaction, according to a theory proposed by Professor Günter Wächtershäuser, is the formation of iron(II) sulfide from hydrogen sulfide released by hydrothermal vents and iron(II) ions dissolved in the waters of a primordial ocean. But iron(II) sulfide, FeS, goes on to further react with H_2S and so produce pyrite – "fool's gold" which is better known to geologists than it is to fools. Its formula is FeS_2, and its formation releases electrons ($FeS + H_2S \rightarrow FeS_2 + 2H^+ + 2e^-$). Electrons are the source of energy found not only in our electric devices, but in our living cells as well. This is so because electrons released by one reaction (or half-reaction to be precise) may be used to "push" another one – they can be used for reduction of organic as well as inorganic substances.

This was a real breakthrough. "There are two alternatives concerning the origin of life: the origin may be heterotrophic or autotrophic," wrote Wächtershäuser in the abstract of his 1990 paper published in the *Proceedings of the National Academy of Science USA*. "The central problem within the theory of an autotrophic origin is the first process of carbon fixation. I here propose the hypothesis that this process is an autocatalytic cycle that can be retrodictively constructed from the extant reductive citric acid cycle by replacing thioesters by thioacids and by assuming that the required reducing power is obtained from the oxidative formation of pyrite (FeS_2)." (The process is called autocatalytic because every member of the cycle reproduces itself by entering it.) The body of this paper presented an amazing scheme for hypothetical prebiotic citric acid cycle comparing it with extant reductive citric acid cycle found to be operative in bacterial and archaeaic domains and, which is of particular importance, in thermophiles thriving on hydrogen sulfide. The two proteins with Fe:S clusters do participate in this metabolic cycle!

Ethica ordine geometrico demonstrata, Ethics Demonstrated in the Manner of Geometry, is the capital work of the seventeenth century Flemish philosopher Baruch de Spinoza. At its center is a very strange notion that rules of proper human behavior can be demonstrated in the form of mathematical theorems, and it is no less unusual to find such reasoning in a modern chemistry paper. Namely, Wächtershäuser, in the best tradition of continental philosophy, developed his theory through five postulates and eight theorems which I have no intention of reproducing here. Suffice to say that this German scientist modified reductive citric acid cycle by "replacing thioesters by thioacids," as mentioned previously, i.e. by replacing alcoholic (–OH) functional groups by thiol ones (–SH), and substituting dozens of cell enzymes with just one catalyst – FeS. In the reductive citric acid cycle, as well as in the oxidative one (which appears in higher organisms, including humans) reducing power is provided by NADH or its phosphate, NADPH ($NADH/NAD^+$ or $NADPH/NADP^+$ system). In his hypothetical prebiotic cycle they are replaced by FeS/H_2S reaction. The key enzyme ferrodoxin, a protein with Fe:S cluster, is substituted with FeS, and the complex coenzyme A (HSCoA) with H_2S. All these reactions occur on a sulfide matrix because "The geochemical setting of the origin of the archaic precursor cycle is an anaerobic aqueous environment, rich in H_2S (or HS^-) and in contact with heavy metal sulfides and pyrite," as states the first postulate of Wächtershäuser's "mathematical" scheme.

But in his 2006 year paper the German scientist took a step further, proposing a scheme for the formation of amino acids and peptides on a heavy metal sulfide catalyst, namely on (Fe/Ni)S surfaces. In this synthesis route carbon monoxide, hydrogen sulfide and ammonia all participate, as gases found in "every" scheme of prebiotic synthesis. The key step in this scheme is formation of carbonyl sulfide (COS), a gaseous substance whose smell I experienced many times in my youth when steam locomotives with big capital letters UNRRA* passed near my house. It may be easily formed from carbon monoxide and hydrogen sulfide ($CO + H_2S \rightarrow COS + 2H^+ + 2e^-$) and the released electrons can be used to convert carbon dioxide into methylthiol ($CO_2 + H_2S + 6H^+ + 6e^- \rightarrow CH_3SH + 2H_2O$). From those substances, COS and CH_3SH – also using sulfide catalyst – alanine, amino acid next to glycine in simplicity, could be formed. And it doesn't end here! The father of iron-sulfur-world theory proposed a cycle for peptide synthesis on (Fe/Ni)S matrices by using carbon monoxide for the propagation of peptide chains. Furthermore, the proposed cycle could also be used for the formation of purines which are a major component of nucleic acids. It is possible that all this took place on sulfide surfaces and membranes** pulled by reaction of pyrite formation, reaction of hydrogen sulfide with iron(II) sulfide, FeS. Some of those reactions are presented in the following table.

But the central problem still remains. FeS particle is just a particle, an entity more akin to contact catalysts found in a chemical factory than to a living cell. To make a catalyst alive it is necessary to confine it in some kind of cell, to jail it – but at the same time taking care not to place it in solitary confinement. In other words, some kind of membrane has to form in order

---

*UNRRA, United Nations Relief and Rehabilitation Administration, a UN institution whose purpose was to render economic aid to European nations immediately after World War II. "What is like living in Yugoslavia?" "Good, good, but it would be better if we had more UNRRA and less UDBA." (UDBA was acronym for the secret police in Communist Yugoslavia.)

**These sulfide particles might have quite a complex composition, including catalytic minerals mackinawite, $Fe(Ni)_{1+x}S$, and greigite, Fe_3S_4. This last mineral is found in clays as well as in hydrothermal vents and has a structure resembling (Fe/Ni)S clusters. Further development of the theory that life began near iron sulfide chimneys came from Michael J. Russell, who studied formation of iron sulfide vesicles and membranes by performing "chemical garden" experiments.

Possible reactions on the (Fe/Ni)S catalyst in the primitive Earth ocean

Name of reaction ... (corresponding to the metabolism of modern organisms)	**... and its equation**
Nitrogen fixation, production of ammonia from free nitrogen (*Rhizobium leguminosarum*)	$N_2 + 3FeS + 3H_2S \rightarrow 3FeS_2 + 2NH_3$
CO fixation, production of pyruvate from carbon monoxide	$3CO + 4H^+ + 4e^- \rightarrow CH_3\text{-}CO\text{-}COOH$ $(3CO + 2FeS + 2H_2S \rightarrow$ $CH_3\text{-}CO\text{-}COOH + 2FeS_2)$
Synthesis of alanine from pyruvate (transamination, all organisms)	$CH_3\text{-}CO\text{-}COOH + 2Fe^{2+} + NH_3 + 2H^+ \rightarrow$ $CH_3\text{-}CH(NH_2)\text{-}COOH + 2Fe^{3+} + H_2O$
Synthesis of methylthiol from carbon dioxide	$CO_2 + 3FeS + 4H_2S \rightarrow$ $CH_3SH + 3FeS_2 + 2H_2O$
Synthesis of activated thioacetic acid (activated acetic acid, acetyl-CoA)	$2CH_3SH + CO \rightarrow CH_3\text{-}CO\text{-}S\text{-}CH_3 + H_2S$
Synthesis of succinic acid, a component of the citric acid cycle	$4CO_2 + 7H_2S + 7FeS \rightarrow$ $(CH_2\text{-}COOH)_2 + 7FeS_2 + 4H_2O$

to divide substances residing inside and outside of the "living" thing. But then again this boundary has to provide some communication as well, a passage of substances between the inside and the outside. And iron-sulfur-world theory has just the solution for this conundrum. The first cells, pioneer organisms, were developed by lipophilization of catalytic surfaces, by smearing them with some kind of oil. And so we come to the next section.

Pioneer organisms

The first organisms, "pioneer organisms," as the father of iron-sulfur-world theory named them, were (Fe/Ni)S particles covered by a "smear," a film of organic compounds. To be more specific, pioneer organisms were composed of a mineral substructure and an organic superstructure immersed in the waters of the first Hadean ocean, and called "volcanic liquid water phase" by chemists. Though iron(II) sulfide was prevalent in the inorganic substructure, there was also some nickel(II) sulfide, as well as iron(II) hydroxide, $Fe(OH)_2$. This last compound was capable of donating electrons by its conversion into iron(III), i.e. $FeO(OH)$, a chemical compound with a long and complicated name though commonly known simply as rust.

The second component of this pioneer organism, its organic superstructure, consisted of organic compounds adsorbed on a mineral substructure. These organics were of numerous diverse kinds and their complexities ranged widely. They were also in a constant state of wild chemical transformation because, as mentioned before, they were effectively bonded to the surface of the catalyst. On the other hand, in contrast to industrial catalysts, pioneer organisms represented a remarkable combination of the three capacities which enabled them to supersede everything else on the planet. They were capable of growing, reproducing and evolving. If we accept this notion that life in primordial broth began with the emergence of an entity with such characteristics, then Wächtershäuser's pioneer organisms were most certainly alive.

But what does it really mean that they possessed the capacities to grow, reproduce and evolve? First of all, they converted carbon monoxide, carbon dioxide and other simple compounds into organic substances of higher complexities. Organic matter gradually accumulated on a mineral substructure. So that is the capacity for growth solved. Catalytic reactions which provided this initial growth then gradually evolved into autocatalytic ones, marking the first step in differentiation of the living from the non-living. This occurred because autocatalytic reactions are defined as those that produce their own catalyst. In other words, they reproduce the initial substance and so exhibit a positive feedback. This is also highly specific to life, for all life processes are controlled by some kind of a positive or negative feedback.

The main problem with chemistry, as we are taught during our chemical education, is that chemical transformations are considered to be well defined chemical reactions. It is quite an unusual occurrence for a classroom experiment to go astray. Chemistry is something happening in test tubes involving reactions of chemicals taken from laboratory shelves. However, there was nothing of the kind happening in those early Earth oceans, in the vicinities of hot water springs enriched with dissolved volcanic gases where the pioneer organisms are assumed most probably to have originated. Abundant differences in their compositions resulted in high rates of variation. This means that some of their autocatalytic cycles were better fit for survival than others and thus went on to develop further, which is just another way of saying that evolutionary processes began to take place.

And there is an even more elaborate description of this process. It stems from the fact that organic molecules were either bonded to the surface of (Fe/Ni)S catalyst or were bonded, by weak intermolecular bonds, to the molecules already attached to it. This resulted in a constant interchange of organic matter between the solution and the inorganic surface.

What is of utmost importance though is that some of the molecules bonded to the catalyst served to enhance its catalytic activity. These molecules were bonded to metal centers, that is to the free valences of iron and nickel, and most tellingly and very much in favor of the chemoautotrophic theory, they were molecules with amino ($-NH_2$) and carboxyl ($-COOH$) groups. In other words, hydroxy acids, such as lactic and citric acid, as well as amino acids, like glycine and alanine, increased the activity of the catalyst, i.e. inorganic substructure of the pioneer organism.

In brief, increased production of a chemical compound on the catalytic surface also acted to increase its activity, which resulted in even more of it being produced. As this compound converted into other compounds, the rate of their formation increased as well. "This means that the autocatalytic feedback effect of the one special product increases the steady-state concentration of each compound in the set of reaction products," wrote Wächtershäuser in his 2006 paper, adding that "the metabolism of the pioneer organism is self-expanding and undergoing a progressive increase of metabolic diversity and molecular complexity, thus exhibiting an ever-expanding avalanche of metabolic evolution." Life did not emerge by some miraculous immaculate formation of the first living cell or a "live" molecule (such as RNA), but by the first autocatalytic reaction on the surface of iron(II) sulfide particles in the depths of some primordial ocean.

The next step in the evolution of pioneer organisms was the formation of a lipid layer on an inorganic substructure (surface lipophilization). Fatty acids, the principal components of lipids, are synthesized by action of "activated acetate," acetyl–CoA. As this constituent of a living cell is essentially a thioester, CH_3–CO–S–CoA, Wächtershäuser proposed that the first lipids were fatty or isoprenoid acids synthesized by thioester condensation. Though such lipids cannot be found in modern organisms anymore, fatty and isoprenoid acids should also be considered as lipids because their molecules, as with molecules of any lipid, may be split into hydrophilic (head) and hydrophobic parts (the rest or tail). It is quite logical to assume that molecules of fatty acids would tend to direct their heads (–COOH) towards the water phase while their tails (composed of long hydrocarbon chains) get anchored to the hydrophobic inorganic substructure. Following this formation of lipid molecules, a bilayer membrane would eventually form spreading itself over the surface of iron sulfide particle.

But then again, this membrane, as the story of the origin of life is now understood, did not manage to spread over the entire surface of the particle. This is because its surface wasn't smooth and uniform but rough and irregular, covered by microscopic concave crevices and cavities, as occurs naturally on surfaces of inorganic materials. The bilayer membrane then covered these crevices forming a lipophilic barrier against their watery surroundings. A semi-cellular structure was formed in this way, for the "embryonic cavern" was enclosed by a bilayer membrane on one side and by the solid surface of the particle on the other.

Catalyzed reactions continued to take place in this embryonic cavern, though the communication with the environment was now limited by the permeability of this newly formed membrane. This "semi-cellular structure" gradually increased in size and complexity with the membrane increasingly enveloping the contents of the embryonic cavern, until it became fully separated and independent of its solid support. And so the first cell was born. (Such a scenario is now considered to be even more probable since it was shown that mineral particles, especially those of montmorillonite, accelerate assembly of fatty acids into vesicles.)

These already established catalytic processes continued in the cell state, though only on smaller FeS particles and FeS clusters. From them the first enzymes evolved. This is why we find iron-sulfur proteins in extant organisms, and see their involvement in the most basic cell processes.

The mystery of nucleic acids

At a first glance the iron-sulfur-world theory looks like nothing more than a new brand of the old Oparin's theory of coacervate droplets. Both theories start at the same point, the synthesis of complex organic compounds, and both then go on to separate these compounds from their surroundings by formation of protocells. Both theories regard life as a system of coupled catalytic reactions. But there are differences, of course. Oparin assumes that catalysis commenced following the production of complex organic compounds and their subsequent confinement in protocells. On the other hand, Wächtershäuser believes that the catalyst, that is FeS particles, already existed before any complex organic compounds were formed, and cellularization began only following the rise of metabolism, that is autocatalytic reactions on FeS particles.

"The essence of life is autocatalysis," I noted in one of my popular-science lectures on the origin of life and iron-sulfur-world theory in particular. This notion was not agreeable to one well informed listener, who quite reasonably commented that the essence of life is in the realm of proteins and nucleic acids or, specifically, in their respective metabolisms. All life on this planet follows the central dogma, DNA → RNA → protein, DNA makes RNA makes protein. A theory which is not able to give an answer to the question of the origin of this genetic machinery can't be regarded as a valid theory of bioneogenesis, which is the scientific term for the origin of life.

Oparin's theory begins with proteins, i.e. organic catalysts. Nucleic acids and genetic machinery associated with them evolved afterwards. Conversely, theories of RNA world, which were presented in the previous chapter, begin with nucleic acids while assuming that proteins appeared later. Curiously, the iron-sulfur-world theory begins with neither proteins nor nucleic acids, but it does present a very strange idea about their evolution, and especially about the origin of genetic machinery.

The proposed metabolic cycle on (Fe/Ni)S particles describing the formation of amino acids and their dimers with subsequent elongation of polypeptide chain, also involves formation of hydantoin, a heterocyclic compound quite similar to bases of nucleic acids. Furthermore, a kind of hybrid between heterocyclic compounds and amino acids is to be found in this reaction scheme, leading to a quite logical hypothesis that the

1 380

great increase in
stromatolite diversity

first biopolymers were neither proteins nor nucleic acids but compounds sharing structural elements of both polymers.

The idea that life arose from such compounds, polymers resembling both proteins and nucleic acids, culminated in the theory of peptide nucleic acids (PNA). In 1991 Eschenmoser and associates proposed prebiotic existence of curious polymers in which ribose-phosphate backbone was replaced by glycine units, and its structure could also be described as branched polypeptide with nucleic bases in its side chains. PNA polymers are even capable of forming stable double helices with nucleic acids, with RNA as well as DNA. However, this greatly differs from polymes envisaged by chemoautrophic theory. Its starting point is polymers resembling an amino-acid molecule bonded to transfer-ribonucleic acid, tRNA.

In all living organisms protein synthesis takes place on ribosomes through a reaction of two amino acid-tRNA molecules (like Gly-tRNA, Ala-tRNA, Ser-tRNA, etc.). Such molecules, pro-tRNAs (ancestors of tRNAs) were bonded to the (Fe/Ni)S catalyst and were produced, like other molecules, by autocatalytic positive feedback. The first assumption is that an amino acid was attached to the end of the pro-tRNA molecule. "Let us further assume that two of these surface-bonded pro-tRNAs became located side by side and attached by base pairing with their loops to a surface-bonded nucleic acid strand as pro-mRNA," wrote Professor Wächtershäuser. The consequence of such an alignment is straightforward: the synthesis of dipeptides, and olygopeptides in later steps, was directed by the sequence of pro-mRNA.

A better base pairing would have led to better protein synthesis, and an indirect, secondary positive catalytic feedback mechanism was thus gradually established. This better base pairing led to better pro-tRNA and pro-mRNA synthesis, the better RNA synthesis enhanced protein synthesis, and the improved protein synthesis produced greater variations which finally led to accelerated development (evolution) of the first "living" system, the pioneer organism. This means that both replication and translation were established jointly, by coevolution: "the evolution of nucleic acid replication and of nucleic acid-catalysed peptide synthesis must have been intrinsically linked." Nucleic acids and proteins did not evolve separately to end up joined together in the course of evolution. On the contrary, genetic machinery evolved as a whole to ever greater levels of fidelity and organic functionality.

This is the final point of the theory which seems to me the most complete and convincing one of all the previously described theories of life origin. It was supported by the discovery of chemoautotrophic hyperthermophiles in the depths of the ocean and other equally exotic places. In the 1970s Thomas Gold, perhaps the most imaginative geophysicist of the last century, proposed his "deep-hot biosphere" model. His findings of filament structures in deep rocks, up to five kilometers below the surface, led to a strange notion that the first organisms on the planet were extremophiles, organisms originated in and consequently adapted to harsh environmental conditions. (Hyperthermophiles are just one type of extremophiles; there are halophiles who live in salty waters of the Dead Sea, acidophiles who dwell in the acid waters of Rio Tinto, xerophiles that resist desiccation, barophiles, alkaliphiles, etc.)

According to Gold, the first organisms' food was hydrogen or methane. These were released by action of hot water on iron (hydrogen) or iron carbides (methane) on deep rocks. In the 1980s Günter Wächtershäuser provided a more elegant theory by introducing hydrogen sulfide as the source of energy and proposing autocatalytic positive feedback on FeS particles. From a geological point of view, these theories are more convincing still. Namely, it is difficult to imagine that life originated and propagated on the surface of a primordial ocean. The UV radiation was too powerful back then – as much as 10^{31} times stronger than today!

Did life come from outer space?

1 290

Mill. yrs from present

7. BURNING ICE CREAM

Perhaps life arose from fire, as a consequence of meteorite and asteroid impacts at the beginning of Earth history. So, in a way, life did originate from space...

Long before the term *team building* found its way into dictionaries, our laboratory boss used to invite the whole staff to his home to celebrate his birthday. I took upon myself the responsibility of buying flowers for his wife, and my colleague Nick's task was to select a bottle of premium wine. Others bought nothing, though they provided the money for purchases. And so all of us would appear at the door of the house where he lived with his family. Our boss' name was quite German - Otto A. Weber, but nobody knew what this A stood for, especially because it is quite uncommon to have two first names in Croatia. It was only after his death in 1994 that we learned what was hiding behind that A. It stood for Adolf; the name Adolf Weber may not be equal in impact to, say, Adolf Hitler – but our Otto Adolf wasn't as unwise to expose his flanks in that fashion, especially not in a Communist country still traumatized by the horrors of Nazi occupation.

In his youth, Otto proved himself a very diligent and industrious scientist. His scientific career was quite straightforward: immediately after graduating at Zagreb University he landed a job at the Institute for Medical Research in the same city, followed by a three-year stipend at Oxford. Later on he moved to Melbourne, Australia for further three years, working at the institute of the same name. But his career was still being held back by two very important factors. The first one is that he was not a member of the Communist Party, which meant that he was not considered fit for any job of real responsibility. To clarify, institutes as well as companies in former Yugoslavia's "self-management socialism" were organized in pretty much the same way as is customary in parliamentary democracy governments. Formally, all power resided in the workers' council, and the director played the role of a prime minister and was thus nominally directly responsible to the workers' council. But there was an

stromatolites decline

important difference. To be deemed fit for election, a director would have to possess "moral and political fitness," which was just another term for membership in the Communist Party. And so the Communist Party controlled the director, and he was doing his best to control the workers' council. To elect someone who was not a Communist to a position of importance would mean that there were no means of controlling him; therefore while our boss was very active in the institute and state administration as well as in WHO and similar international organizations, he has never been elected to the level of a director or some similar position.

The second weakness of Professor Weber was his coronary medical condition. Following a very risky open heart surgery he decided to live in peace and delegated all his laboratory responsibilities to younger colleagues. Everyone respected him immensely and he addressed everyone and was addressed by everyone by *Vi* (a Croatian equivalent to German *Sie* or addressing someone by Mr or Mrs). Possibly inspired by his days in Oxford, he organized regular tea parties at ten o'clock to have a chat with his coworkers, but that was the extent of his direct involvement with staff affairs. And so we remember him mostly by his birthday parties.

You could say he was a party animal, but of a particular, quite modest kind. He enjoyed barbecue though he approached it as a kind of chemical experiment; nobody knew what would end up in the fire. And at the end of one of his birthday parties he came into the dining room with an ice cream enveloped in fire. Of course, we were all familiar with this mode of food preparation, or to be more precise, food presentation, *Cerises flambées, Poires flambées* or *Bananes flambées Martinique*, as one can discover in the *L' Art Culinaire Moderne* composed by the great Henri-Paul Pellaprat. The French chef used strong alcoholic beverage (rum) on fruits only (cherries, pears, or bananas), but who on earth would have come to an idea to do the same with ice cream, ice actually?

This used to present not only a great spectacle, but a nice scientific experiment in itself as well. At the time, many of us were engaged in construction of a calorimeter to be used for medical research. The basic idea was to build an instrument which would measure minute changes of temperature taking place during chemical reactions. It was designed to measure heat released by just a few drops of reactants. The central part of the design consisted of a copper block which had to maintain constant temperature despite variations in the surroundings. At first it seemed impossible to limit temperature oscillations inside the block within 1 µK when the temperature of the ambient air oscillated as much as 0.1 K or 100,000 µK. But it worked. It worked because it takes time for heat to transfer from a hotter body to a colder one. And the same holds true for ice cream.

There are at least as many air bubbles in ice cream as there are in snow, which makes both of them heat insulators of the similar quality as wool: snow cover keeps corn fields warm during winter periods, and the traditional Inuit feel quite comfortable in their igloos. This is why heat released by burning alcohol cannot do much harm to ice. The result is ice cream roasted at the surface, though cold and solid below it.

If you were to repeat the ice-cream experiment of Professor Weber, the last of our chemical cooks, you'd be more inclined to believe that the same regularly happens with meteorites. They also come to Earth with surfaces melted while their interiors remain largely intact. And so it is that with them life began...

Workings of falling bodies

If you open any one of the many books on the history of early Earth, you will probably find a paradoxical fact, expressed with more or less astonishment by the author, that life on our planet emerged very quickly, perhaps in just a few tens of millions of years, even though the process leading to its origin is thought to be rather complex and obscure. Life began soon after the initial period of intense collisions between Earth and various huge space bodies, asteroids and even planetesimals, finally ceased. This leads us to a logical assumption that life repeatedly "kept beginning" during the Earth's early history and was then continuously wiped away by sterilizing power of large-sized impactors arriving from outer space. It is even possible that hyperthermophiles were not the first organisms, as described in the last chapter, but merely the only survivors of these cosmic cataclysms.* It is not

*It was estimated that the kinetic energy of 440-km asteroid (1.3×10^{20} kg) with a velocity of 17 km/s would be quite sufficient to evaporate the entire Earth's ocean. On the other hand, microbes easily adapt to the temperature of their surroundings. Those living in water and soil have the optimum growth temperature 22 – 28 °C, but all pathogen microorganisms grow best at body temperature, 37 °C.

1 200

the oldest fossils
of red algae

even essential for our argument to presume that some terrestrial life did manage to survive sterilization by impactors, for example at some remote spot in the ocean. According to results of computer simulations, it is quite possible that rocks ejected by these impactors returned to Earth several thousands of years later carrying living cells within them. Life perhaps did not come to Earth from Mars, as was suggested in the second chapter, but instead returned to its cradle after a long space voyage. Calculations have shown that about one percent of the ejected material did find its way back to Earth. If just one cell per kilogram of fallen rocks did survive, life on the planet might have started again.

But there is another link between large-sized impactors and the chemo-autotrophic or iron-sulfur-world theory. Perhaps this theory has nothing to do with undersea "black smokers," volcanoes and their fumes dissolved in hot sea water. Hydrothermal circulation may well have been necessary for the origin of life, but it did not come into being by volcanic activity, but by the very act of a big rock falling from space.

In the second chapter we mentioned that collision of Earth with another cosmic body, even of a moderate size, releases tremendous energy. A sizable portion of this kinetic energy is used up in melting and even evaporating rocks, as well as the body of the impactor itself, while another portion is transformed into a shock wave producing earthquakes and tsunamis. It is also expended in creating cracks in the rocks near the crater. Water can easily find its way into these cracks – turning them into fully-functioning hydrothermal systems.

In 2006 American scientist Charles S. Cockell presented four general requirements which had to be fulfilled by any theory of the origin of life which aspires to be convincing. The first requirement is to propose a source of energy for chemical synthesis of small as well as of big molecules (polymers). The second one is to provide a suitable mechanism for concentration of reactants at a place which would facilitate chemical changes. The third one is the existence of a suitable catalyst. And the last, fourth requirement, is to provide a geochemical environment in which these changes can take place over the long, long periods of time needed for the development of the first protocells. Could the craters made by large-sized impactors be such places?

If collisions of our planet with bodies of adequate size were enough to boil its ocean "n = 1, 2, 3 ? times" (as I found in one chronological table), it is then clear that collisions with smaller bodies should have been enough to give rise to some cracked rocks around the crater filled with hot water over a very long period.

Cockell named 16 big impactor craters, with diameters ranging from 870 meters (Wolfe Creek, Australia) to whopping 300 kilometers (Vredefort, South Africa), and from 52,000 (Lonar, India) to 2.023 billion years old (Vredefort), wryly noting that "ages are correct at the time of writing." A zone of fractured rocks extends around each crater. One such zone around the 52 km diameter Siljan crater in Sweden (361 mill. years old) extends five kilometers below the crater floor and encompasses a volume of more than a thousand cubic kilometers! And we know that hot water did circulate through these cracks. But – for how long?

How long does it take to cool hundreds of cubic kilometers of rocks from the initial temperature of 650 – 700 °C, as was estimated for the hydrothermal system of the 24 km diameter Haughton crater in Canada? About 10,000 years. Is this enough time for the formation of autocatalytic FeS or similar particles and their development into something alive? Hardly, but the 250 km diameter Sudbury impact structure in Canada sustained its hydrothermal system for more than two million years. Apparently, this would be quite enough if we accepted the notion that life emerged on the surface of Earth 4 – 3.7 billion years ago "$n = 1, 2, 3$? times."

"What might have been is an abstraction/ Remaining a perpetual possibility/ Only in a world of speculation," wrote T. S. Eliot, my favorite poet. To place the beginning of life at four billion years before the present, supposing that life began and ceased to exist many times in the half billion years period before the first evidence of it emerged in the rocks, is merely a speculation and not scientific fact. But the same cannot be said for chemical reactions.

A hydrothermal system of an asteroid crater does not simply consist of some hot water plus cracked rocks. The initial heat as well as subsequent action of hot water worked to alter many minerals, as much as 25%, as was discovered in cracked rocks around craters. Hot water reacts with alumosilicates and glasses produced by the shock wave, converting them into zeolites and clays. Among them the most important one is montmorillonite, a well-known mineral catalyst for synthesis of proteins and nucleic acids, quite instrumental in assembling vesicules as well. Heavy metals, especially iron, dissolve in hot water and are precipitated as sulfides later on. A crater's hydrothermal system is a huge chemical reactor, holding a volume of many hundreds of cubic kilometers. Something alive has to emerge from it – after a few million years.

Careful chemical research designed to reproduce events in the hydro-thermal system of a high-sized impactor supports this supposition. It was demonstrated that at 300 ºC a mixture of hydrogen, carbon dioxide and carbon mononoxide was converted into alkanoic acids, and at a lower temperature of 200 ºC, alkyl formates were produced. According to calculations, the optimal temperature for organic synthesis in hot water is 150 – 250 ºC, which is quite a bit lower than the temperature of water spewed by "black smokers." Furthermore, water found in the cracks is weakly alkaline to near-neutral (pH = 6 – 8), which makes it suitable for precipitation of iron(II) sulfide and its use as a catalyst. These events proposed by the chemoautotrophic theory seem even more probable when placed in the cracks of an asteroid crater than on the bottom of the ocean. And there is yet another point in favor of such a scenario.

Impact crater structures are not homogeneous, neither in time nor in space. At high temperatures near the bottom of the crater and not for too long after the impact, fractured rocks generated hydrogen and other low-molecular gases. At lower temperatures and at lower depths, low-molecular compounds formed first, followed by high-molecular ones. Water current then transferred them closer to the surface where, at much lower temperatures, the first protocells might have been formed. Exposed to light, the first living things evolved into photosynthetic (or photoautotrophic) organisms, thus starting life in the modern sense of the word.

Circulation of water in this peculiar chemical reactor was provided by gradual cooling, and this water carried "food" for reactions which gradually grew in complexity. This was a true realization of Darwin's "warm little pond" (with slight exceptions that it wasn't warm but hot, and not at all little but quite big instead) and the most convincing explanation of how life originated in magmatic rocks, as originally proposed by Thomas Gold.

But there is something missing from this story (or history). Life has its beginnings in a hot environment, starting with simple carbon compounds released by reactions of hot rocks with hot water, very close to what Oparin proposed at the beginning of the last century. This theory proposes that action of heat and water converted all organic matter into the simplest of compounds, such as carbon monoxide. From these compounds, by action of catalysts, more complex organics were eventually synthetized, a hypothesis quite in line with iron-sulfur-world theory. But is this the end of the story?

Chemistry of fallen cosmic bodies is not as simple as that though.

Chill beginnings

At the end of ten long years of wandering through empty space, *Rosetta* arrived at its destination: in November 2014 it reached a comet marked 67P and named Churyumov-Gerasimenko, or 67P/C-G for short, after its Russian discoverers. *Rosetta* began to orbit the comet and landed its 100 kg probe *Philae* on it (at the moment of landing its weight was the equivalent to the weight of two grams on the surface of Earth).

Everything about this event was fascinating. How could it possibly be that something made by human hands endured ten years of arduous voyage through cosmic vacuum and, even more impressively, the damaging effects of solar radiation? How is it possible to find a speck of rock millions of kilometers from Earth and land on it? But to me the most fascinating aspect of this endeavor was that the landing on the comet and the subsequent collecting of ground samples was all accomplished by crafty utilization of weak gravitational forces; and it is even more fascinating that all the loose particles which comprise the comet are kept together by the same, equally weak forces. Comets are not simple "dirty snowballs," as they were visualized thirty or forty years ago, but dusty rocks (or, to be precise, rough mixtures of water and silicates).

But no matter how weak the force of gravity is, how small comet bodies are, how their tails are composed of matter less dense than the best laboratory produced "vacuum," the laws of nature still do not lose their potency. Laws of physics and chemistry are always the same irrespective of the amount of matter involved. It is possible to calculate attraction of two dust particles in the same manner as we calculate attraction forces of two planets, two stars or two galaxies by using the same Newton's equation of gravity and do so at nearly the same level of precision. Likewise, it is possible to study chemical processes on comets despite minute concentrations of reactants. Furthermore, we can determine these reactants, their reactions and their products – and that is a very valuable lesson in chemistry. For in chemistry, not a thing is ever just "nothing;" 10^{-5} mg of type A botulinum toxin, which is the quantity estimated to be sufficient to kill an adult human, has the same practical effect as seven kilograms of water (water *is* also a poison!).

Rosetta wasn't the first space probe aimed at a comet. In 1986 the *Giotto* probe made by the same institution, European Space Agency (ESA), approached to as close as 600 km to the legendary Halley comet and took pictures of it. In 2001 NASA probe *Deep Space* did the same with the comet 19P/Borelley, crossing its orbit at the distance of 3000 km. In 2005 *Deep Impact* came close to 9P/Tempel and launched a 400 kg projectile at it. The impact created a cloud of dust around the comet, and its composition was analyzed by the probe's instruments. Seven years later, in 2012, *Stardust* succeeded in taking pictures of the comet, showing a clearly visible crater produced by the *Deep Impact* projectile. In 2004, during its first flyby the *Stardust* spacecraft did something even more spectacular: it came close to the comet 81P/Wild to take dust samples from its tail and then returned them to Earth. In them scientists found many organic compounds, including glycine, the simplest amino acid.

Such evidence is fascinating, and doubtlessly convincing, but it is not the first of the kind. In an old textbook on astronomy, published in 1974, I found that the nucleus of a comet is composed of water, ammonia, methane and carbon dioxide which were converted into a range of ions and radicals (CO^+, N^+_2, CH, OH, NH_2, etc.) by solar radiation. Ions and radicals are very reactive and they react with each other producing more complex organics. A comet's coma (comets are composed of three parts: a *coma*, the glowing head, its *nucleus*, and 10 – 100 million km long *tail*) is actually a very low-density gas. It contains only up to 10^5 – 10^{10} molecules per cubic centimeter (compare it to 2.7×10^{19} molecules in a cubic centimeter of air) but still some chemical transformations, albeit at a much slower rate,* do take place. Time doesn't really matter with comets however. Within their lives, spanning billions of years, anything might happen.

Comets are rare and spectacular celestial events, but as astronomical events they are not that infrequent at all. Our solar system is not composed of "seven planets plus debris," as we believed until the end of the last century. Instead, it is actually a huge cloud of rarefied gas and dust particles,

---

*The rate of chemical reaction is roughly proportional to the product of reactant concentrations; reactions of radicals in a comet's tail are 10^{18} – 10^{34} times slower than in the air.

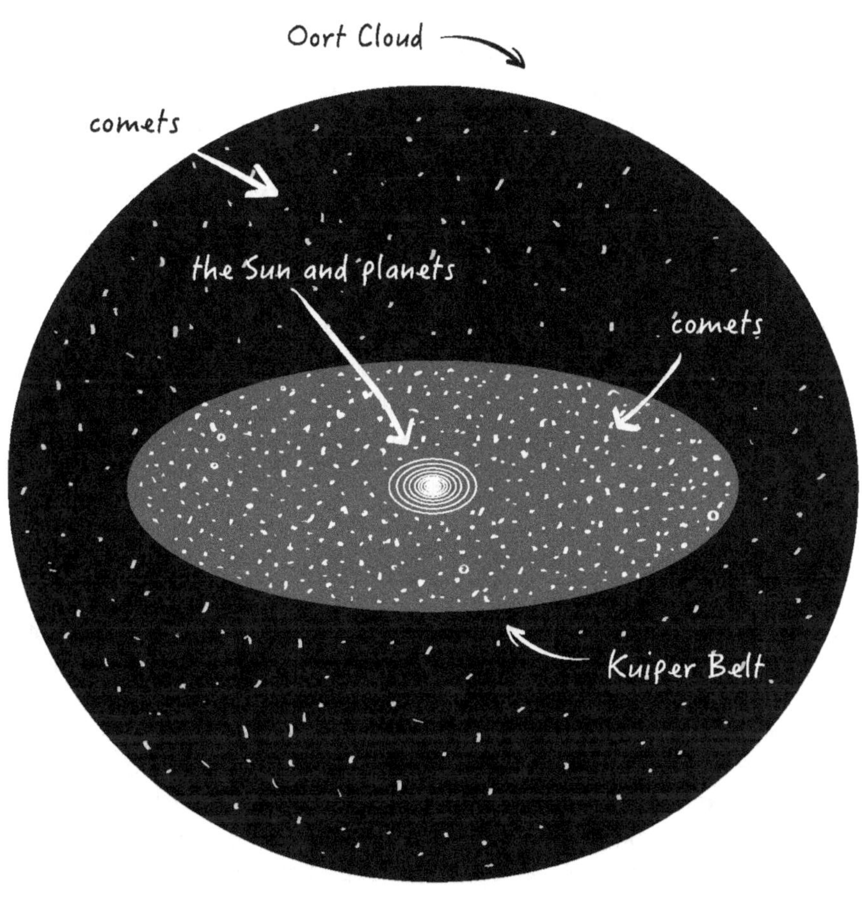

Our Solar System is composed of the Sun, the planets, and various assorted debris – but this "debris" is actually by far the largest part of the whole, comprised of billions of comets and planetoids situated in the Oort Cloud and Kuiper Belt.

1 020

leftovers from the formation of planets. The inner Solar System is enveloped by the Kuiper Belt which extends from the orbit of Pluto outwards. It is estimated that there are 10 billion comets or asteroids in the belt, 35,000 of which might be larger than 100 kilometers. At a much greater distance, about a third of the way to the nearest star, we find the Oort Cloud, a home for perhaps as much as a trillion comets. Approximately every 40 million years this cloud becomes perturbed and hundreds of comets get ejected into the inner Solar System. It is not difficult to calculate that there were about a 100 such perturbations in the 4.5-billion years of Earth history, flinging tens of thousands of comets into crossing the Earth's orbit. Many of them did end up colliding with Earth, with catastrophic consequences (the latest and the most famous of these collisions was the impact of a comet nucleus near the Tunguska River in Siberia, 1908). And there is more. It is quite possible that similar clouds of comets may be found orbiting the other billion (10^9) Sun-like stars in the galaxy, and that these comets may possibly get ejected into other planetary systems, including our own.

Comets have always provoked horror and awe among humans; in the Dark Ages it was believed that comets brought the plague and heralded the end of the world – making them a veritable symbol of death. Comets may turn out not to be heralds of death though, but heralds of life, or – to be more precise – they may herald both because life and death are intimately connected. Death is a natural consequence of life, and life is a natural consequence of death, if you'll permit us a brief digression into philosophical ruminations. This truth is reflected in the Indian Pantheon, especially in the form of god Shiva, "the Destroyer," but also in a theory posited by an Indian astronomer named Chandra Wickramasinghe, a professor at Cardiff University. He elaborated on his predecessor Fred Hoyle's panspermia hypothesis and developed an interesting new theory. "But prior to life being generated anywhere, primordial comets would provide trillions of 'warm little ponds' replete with water, organics and nutrients, their huge numbers diminishing vastly the improbability hurdle for life to originate," he wrote in 2009 on supposition that some of the comets contain, as a result of warming due to radioactive decay, a quantity of liquid water hiding deep within their interior. "Together with Janaki Wickramasinghe and Bill Napier, I have argued that a single primordial comet of this kind will be favored over all the shallow ponds and edges of oceans on Earth by a factor 10^4, taking into account the total clay surface area for catalytic reactions as a timescale of persistence in each scenario."

<table>
<tr><td>1 000</td><td>990</td></tr>
</table>

formation of
super-continent
Rodinia

(Of course, they assumed that hot water in comet interiors transformed silicates into catalytic clays.)

Contrary to what the Indian astronomer claimed, it is difficult to believe that life emerged on as desolate a place as a comet; just look at the pictures provided by the lander *Philae*. On the other hand, it is quite a bit more believable that comets did provide our planet with many complex and important organics which would make the beginning of life possible.

Water, methane, hydrogen cyanide, formaldehyde, methylcyanide, glycine...* Molecules of all of these and many other compounds have already been detected on comets. But what about those which are not detected yet? Glycine could be polymerized into peptides, and formaldehyde is a well-known precursor of formose reaction, leading to simple sugars and other organic compounds. If these reactions didn't take place on the comets, they could have possibly happened on Earth when comet-synthesized organics were mixed with waters of the first ocean. These compounds then triggered a succession of events which eventually led to the emergence of life on this planet.

But there is another way, another possible cosmic connection. Perhaps life did begin only after meteors and asteroids discharged their organics. But then again these chemicals happened to be of quite a different kind.

Searching for the ultimate root

The problem with the theory of evolution and the theory of the origin of life in particular, is that many links in the chain are still missing. While it is quite evident that life is based on proteins and nucleic acids, it does appear that they are a later invention; it is very probable that they developed from a common ancestor. Could their ancestors have been "peptide nucleic acids" (PNA)? Or maybe something similar?

This reminds me of a fictional problem concerning reconstruction of the history of the tram (electric street-car). Let us assume that we are familiar

---

*From the simple oxides and hydrides (CO, H_2O, NH_3 and CH_4) a number of organics were synthesized; alcohols, $(CH_3)_2CHOH$ and $(CH_2OH)_2$, aldehydes and ketones (H_2CO, CH_3CHO, C_2H_5CHO, $(CH_3)_2CO$, $HOCH_2CHO$), cyanides (HCN, CH_3CN), isocyanates ($HNCO$, CH_3NCO), amines and amides (CH_3NH_2, $C_2H_5NH_2$, $HCONH_2$, CH_3CONH_2) and even the simplest amino acid, glycine, NH_2CH_2COOH. *Philae* found 16 organic compounds, seven of which were detected for the first time.

with every component of a street-car, but also that we don't know any-thing about its history. Did it evolve from electric motor? Or from its rails? Well the answer is from neither – it evolved from the omnibus and the horse cart. If we carefully study street vehicles, we will soon notice that they all have something in common, the wheel. The wheel is the common ancestor of all vehicles, and the ancestor of the wheel is rolling timber.

And where is the equivalent of the wheel in living beings, what is the rolling timber of life?

In the nineteenth century chemists began to notice that some groups of atoms remain conserved in chemical transformations. When Liebig and Wöhler carried through a large number of reactions with "oil of bitter al-monds" (benzaldehyde) they found that C_7H_5O, a group of atoms, remained unchanged, and accordingly they marked it as Bz (benzoyl). Consequently they wrote compounds derived from their "oil" as BzH, BzCl and $BzNH_2$. Later on they found many similar groups of atoms, called radicals, and con-cluded their research in 1837 with a triumphant announcement: "In mineral [inorganic] chemistry the radicals are simple; in organic chemistry, the radi-cals are compound; that is all the difference. The laws of combination and of reaction are otherwise the same in these two branches of chemistry."

Laws of organic chemistry are not at all as simple as the two German chemists imagined; the analogy between NaCl and BzCl is false, but some-thing in their theory is not very far from the truth. Organic molecules re-ally do have a "skeleton," a component which mostly remains unaltered and it could be conceivably seen as the "root" (Lat. *radix*) of a molecule. Even today chemists use R in their formulas to mark their "radical" parts, although this radical theory of chemical combination is now firmly rele-gated to history.

"Radicals" are obviously found in biological molecules, but what is less obvious is the story they tell. Heterocyclic rings of DNA and RNA nucleo-bases are "radicals," as are the aromatic rings in side chains of some amino acids. Aromatic rings participate in enzymatic reactions, as mentioned in the fifth chapter, suggesting that they are remnants of an RNA world. The first living things were RNA molecules which further evolved into molecules of DNA and proteins, as claimed by the RNA-world theory. Nevertheless, the evolution tree of aromatic rings might still boast of roots reaching much deeper than nucleic acids and their bases.

The first living cell was formed along with the first membrane. A cell membrane is composed of lipids, and lipids are amphiphilic compounds. This means that their molecules have two sides, a hydrophilic and a lipophilic one. But amphiphilicity is not restricted to lipids only; it is a property of many cellular, subcellular and molecular structures. DNA molecule has a hydrophilic core (base pairing) enveloped first by a lipophilic (base stacking) and then a hydrophilic envelope, composed of sugar and phosphate moieties in direct contact with water. Proceeding further along this line of thought, it appears that the original aromatic compounds, composed of one, two or many benzene rings, evolved by gradual hydrophilisation into nucleic bases, nucleic acids, proteins and all kinds of organics found in living beings.

Such reasoning is more appropriate to Greek philosophy than to modern science. Science requires proof.

And the proof we are looking for may be coming at us from space, as a flaming meteorite!

Amino acids from the sky

In 1969, as many laboratories were preparing to receive their first *Apollo* samples from the Moon expecting to find something "alive" in them (i.e. minute amounts of organics), two big meteorites happened to fall on Earth. The first one reached the surface of the planet on February 8th in Mexico near the village of Allende. The second one fell in Australia on September 28th, near Murchison, a small riverside village in Victoria. And that is how they got their names, as is common in the meteorite business, corresponding to the names of those otherwise quite unremarkable places.

But their chemical composition did turn out to be remarkable. These two meteorites were classified as carbonaceous chondrites, Type III (Allende) and Type II (Murchison), meaning meteorites that are notable for their high content of carbon compounds (1.5 – 4%). From this, Wiik's classification of carbonaceous chondrites, it appears that Murchison has somewhat higher contents of carbon and water than Allende.

Amino acids are the building blocks of life, and they are the most important compounds in the story of life's origin, or at least so they were before Fox and Miller's experimental findings got overshadowed by RNA-world and similar theories. Analysis of Allende samples uncovered many of those acids, though their distribution wasn't uniform. The fact that they were a hundred times more abundant on the surface than inside the meteorite body strongly supported the Earth contamination hypothesis. Conversely, at depths of more than a centimeter below the surface their concentrations dropped to 0.2 – 0.5 nmol/g (about 10 ppb), which is comparable to amino acid content of *Apollo* lunar samples (0.06 – 0.5 nmol/g). This abundance of amino acids was as expected, with the majority consisting of glycine (40 – 60%) and alanine (12 – 18%), the simplest of amino acids. However, a noticeable amount – more than one percent – of polar amino acids (serine, aspartic and glutamic acid) was also found.

And Murchison was even more abundant in amino acids. More than 70 of them were found and altogether they exceeded 60 ppm or about 60 mg per kilogram. Of these amino acids, five were found from which a protein molecule could not be built. They have a hydrogen atom next to a carboxyl group replaced with a methyl group, like $C_2H_5C(CH_3)(NH_2)COOH$, which is the simplest one's formula. Their group name is α-methyl amino acids. But it doesn't end there!

All naturally occurring amino acids, except glycine, have the same L chirality. However, amino acids created abiogenically (as those found in the meteorites) are racemates, mixtures of equal parts of L and D isomers. But this is not the case with α-mehyl amino acids found in Murchison. They have one of the isomers, denoted as (*S*), in excess, as much as 15.2% in the case of the mentioned amino acid. But please, don't be confused with L,D and *R,S* symbols! Enantiomers may be marked L or D, to compare them with the simplest sugar glyceraldehyde, but also by using the *R/S* system or CIP rules proposed by two English (Cahn, Ingold) and one Croatian chemist (Prelog). "Each of us, being of sound mind, does freely agree: 1. to observe the aforesaid RULES in all such cases as they may be applicable, and 2. to refrain from using other rules as may have been or as may be proposed by other powers" was written by hand on a sheet of paper dated "Friday, 13th May, 1966" and entitled "Buergenstock Declaration." This "declaration" by the three (C, I and P) chemists further states "that any infringement of this Agreement shall be punished" either by drinking of "1 (one) glass of liquid refreshment to be supplied at the cost of the infringer" or by reading their common paper on the new terminology.

I don't know if any of the three CIP-chemists faced the "punishment", but many of their colleagues should have if they did put their signatures on their declaration. More than half a century wasn't enough for the old L/D system to get entirely replaced with the new one, especially in biochemistry. But let us return to our story.

In contrast to amino acids ordinarily found in meteorites, α-methyl amino acids have an excess of the (*S*) over the (*R*) isomer. And this is only the beginning. In presence of Cu^{2+} as a catalyst, when reacting with α-keto acids, they exhibit a chiral transfer. This means that a small excess of the (*S*) isomer of α-methyl amino acid produces a small excess of the L isomer of synthesized α-amino acids (or their (*S*) isomers, according to CIP terminology). These synthesized L-amino acids are catalysts for the formose reaction, producing an excess of D-sugars, namely, sugars with the same configuration as of those found in all living beings.

If nothing else, chirality came from the sky.

Tarlike origin

Along with amino acids, a large number of other organics were found in carbonaceous chondrites. It is quite impossible to list all of them here, or anywhere else, because high-resolution mass spectroscopy has shown that their number exceeds one million! It would be quite sufficient to say that all classes of compounds were found in the Murchison meteorite, from alcohols, aldehydes and ketones, α-hydroxycarboxylic acids, amines, to molecules akin to those of sugars, and nucleic bases (uracil, thymine)... 683 positively identified organics in all. But they were found in very minute quantities, varying from 0.05 – 0.5 ppm (*N*-heterocycles) to 332 ppm (monocarboxylic acids). All of them put together accounted for just about 6% of the organic matter. The remaining 94% of organics is something which is vaguely described chemically as "macromolecular material," polycyclic aromatic hydrocarbons (PAHs) or as, rather practically named, "insoluble organic material" (IOM) or kerogen.*

In a way, this presented a paradoxical situation where scientists focused entirely on minute ingredients of meteorites, seemingly having no interest in their main bulk, their kerogen ("a material akin to wax"). Structure of kerogen in many ways resembles the structure of ordinary PAHs (naphthalene, anthra-

---

*Meteorites are also rich in schreibersite, a mineral which, by reaction with water, easily converts into pyrophosphates and other phosphorylating agents, necessary for nucleic acid synthesis.

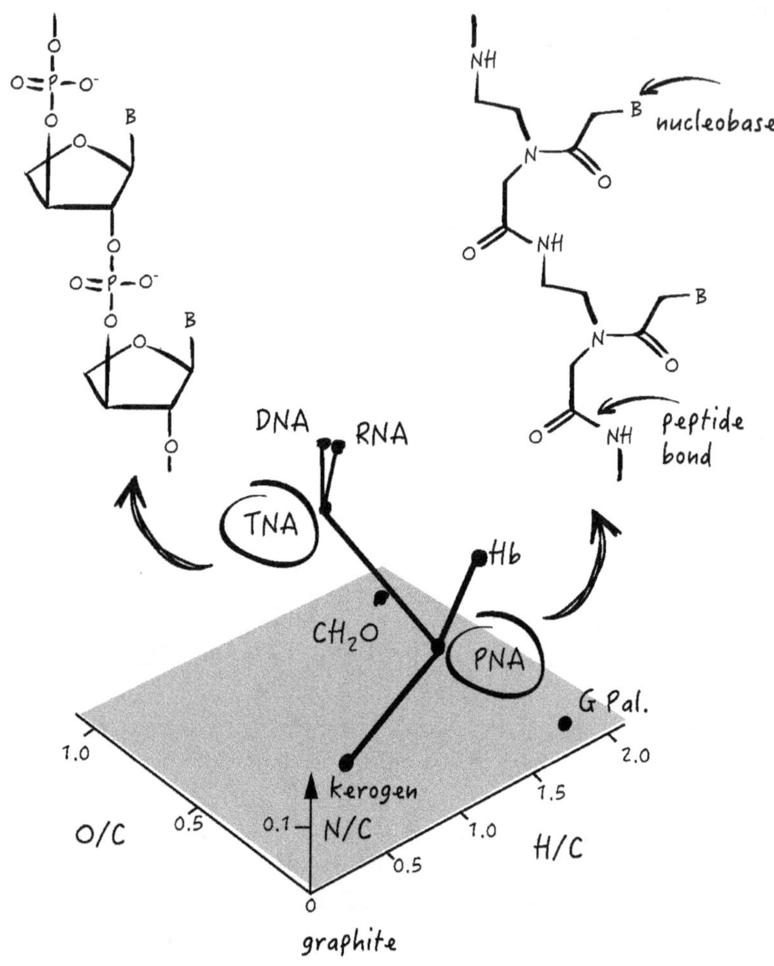

Evolution tree of biomolecules originating from meteoritic kerogen, $C_{100}H_{46}N_{10}O_{15}S_{4.5}$, in H/C, O/C, and N/C coordinates. Through enrichment with oxygen kerogen evolved into lipids (e.g. glycerol tripalmitate, $C_{51}H_{98}O_6$), and saccharides, CH_2O. Enrichment with nitrogen transformed kerogen into peptide nucleic acid, PNA, from which both proteins (e.g. hemoglobin, $C_{2952}H_{4664}N_{812}O_{832}S_8Fe_4$) and nucleic acids presumably evolved. PNA $(C_{14}H_{15}N_7O_2)_n$, as well as the common ancestor of RNA and DNA, threose nucleic acid, TNA $(C_7H_8N_2O_5P)_n$, are here assumed as containing only one "nucleobase" – imidazole.

cene, coronene, and so forth), but in contrast to them, kerogen is insoluble in any solvent, as is suggested by the acronym IOM. It is composed of much larger and more complex molecules whose structure is that of a three-dimensional polymer containing aromatic as well as aliphatic moieties and, contrary to ordinary PAHs, it also contains oxygen, nitrogen and sulfur atoms. Its molecular formula cannot be determined, but it is possible to express its elemental composition as $C_{100}H_{46}N_{10}O_{15}S_{4.5}$. Of all terrestrial materials, anthracite coal may be the closest to it in structure and composition.

And this presents us with another problem. Since meteoritic kerogen cannot be dissolved, we obviously cannot use its solution for proper chemical analysis. Crystalizing it is also out of the question, and if a substance can't be obtained in a crystal state, its full three-dimensional structure cannot be determined. Kerogen may be decomposed by destructive distillation, and its products may be further analyzed, but nothing very valuable can be learned from such analysis. But we can still learn *something*.

NMR spectroscopy revealed the presence of aromatic carbon (PAH), as well as $R-CH_2-R'$, $R-CH_3$, $R-CO-R'$, CH_nO and $R-COOH$ groups of atoms in kerogen. From FTIR spectra it was possible to calculate the ratio of functional groups and other molecular fragments. For Murchison meteorite CH_2/CH_3 groups ratio is about 1.5, ratio $(CH_3+CH_2)/C=C \approx 0.9$, and $C=O/C=C \approx 0.8$. From this data myriads of chemical formulas could be drawn! And what is more, a multitude of molecules of this composition should be present in meteorites. They may be described as polycyclic aromatic hydrocarbons modified by aliphatic chains and carbonyl groups. They may also be viewed as substances similar to products of formaldehyde polymerization ("browning") – which supports the theory that they were produced by a formose-like reaction.

But let us travel back in time, to the very beginning when "Spirit of God was hovering over the face of the waters." Carbonaceous meteorite fell into the ocean. What happened next?

Carbonaceous chondrites are silicate meteorites. This means that they gradually began to decompose, by a process known as weathering, into clays, especially montmorillonite. But what happened with their kerogen? Released from its rock matrix it formed a tarlike layer on the ocean surface. "Meteorite tar" was exposed to UV radiation, and radiation, as usual, effected many changes in its structure. Ions and radicals then attacked the substance, especially its aromatic parts, continually transforming it into something which is quite unlike tar. This original organic material was gradually enriched with carboxyl, hydroxyl, amino and many other groups. The process of weathering began – but now it was organic material that was being weathered.*

"What we tried to stress in our paper is that you have to meet all the requirements at once," said Pascale Ehrenfreund, professor at Leiden University, Netherlands. "You can't have one compound to assemble material, and then add something else later on to do another function. They have to be combined from the beginning – life needs to have an identity, it needs energy, it needs to be able to reproduce and evolve. That's why PAHs are potentially so powerful, because with those aromatic compounds you can fulfill all three functions at the same time."

Everything is possible with PAHs or IOM! By the process of weathering they could evolve into biopolymers like PNA, and further on into proteins and nucleic acids, which could be regarded as kerogen enriched with non-carbon atoms – compare $C_{100}H_{46}N_{10}O_{15}S_{4.5}$ with the formula of DNA, written in the same way, $C_{100}H_{114}N_{39}O_{70}P_{10}$. There was nothing that required polymerization, because IOM was polymerized already. But how did IOM turn into nucleic acids?

DNA and RNA molecules consist of aromatic rings bonded to a cyclic ribose or deoxyribose moiety. It is a well-known structure. However, imagine polymerized benzylnaphthalene after undergoing a weathering process. Fused rings of naphthalene evolve into purine nucleobase, and phenyl ring of benzyl group attached to naphthalene moiety transform, after being enriched with hydrogen and oxygen atoms, into furanoside ring of ribose. The wheel of life is an aromatic ring, and the rolling log is actually meteoritic kerogen!

"Polycyclic aromatic hydrocarbons can be used to build up primitive membrane structures," wrote Professor Ehrenfreund. It is difficult, nearly impossible to reconstruct the weathering processes of meteoritic IOM in a test tube, but experiments with formaldehyde polymers managed to fill the

*Similar to this hypothetical process is the decomposition of wood lignin to humic acids, the major constituents of humus. Lignin is also a complex aromatic compound as is meteoritic kerogen. Its approximate formula is $(C_{31}H_{34}O_{11})_n$. Weathering of silicate rocks and plant residues creates soil, weathering of meteoritic kerogen created life!

750

the oldest fossils of
"animals" (protozoa)

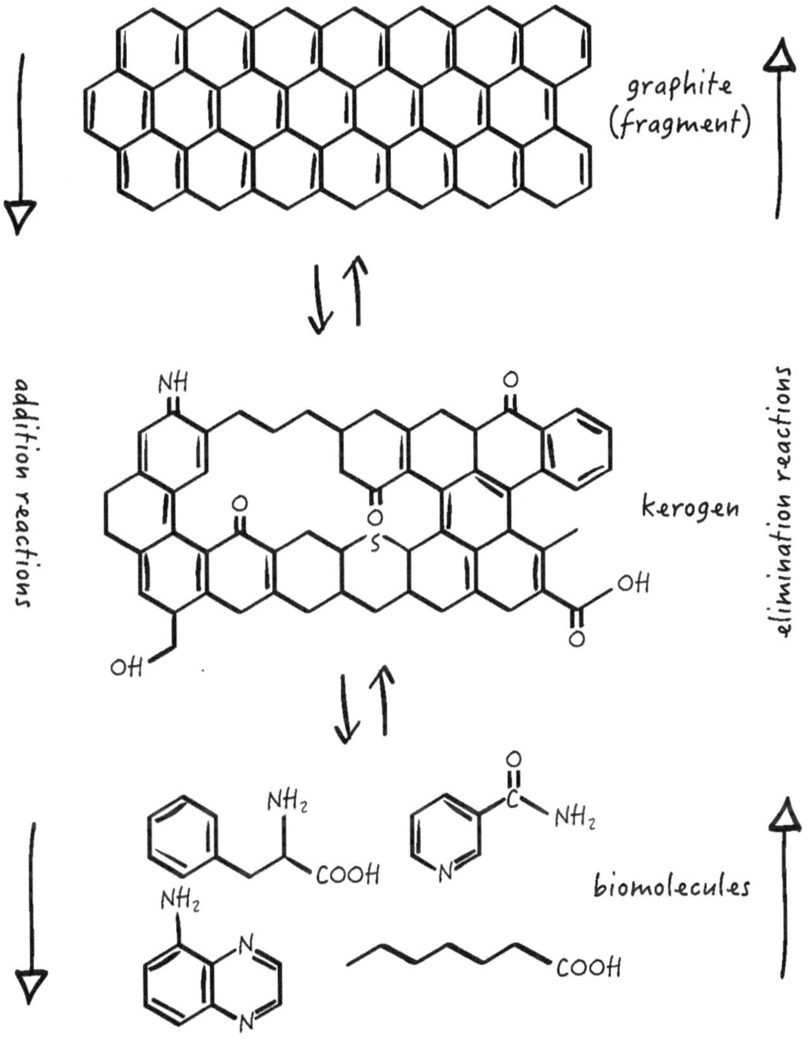

The proposed evolution of meteoritic kerogen into biological compounds is in actuality a reverse process of its geological formation by fossilization of organic material. From a purely chemical standpoint, it has to be noted that elimination is a reverse process of addition.

720

Cryogenian (Sturian and Marinoan) glaciation (Snowball Earth) gap nicely. In an experiment performed in 2013, a solution of formaldehyde (CH_2O) and calcium hydroxide was sealed into ampoules and heated for 72 hours at temperatures between 90 and 250 ºC. Products thus obtained were very similar to meteoritic kerogen. And what is most interesting, those polymers commonly exist in the form of hollow spheres ranging from less than 1 μm up to 10 μm in diameter – exactly the same size as bacteria!

Were these spherules the first living cells?

"PAHs also can be photosensitizers, because they can do a charge transfer between plus and minus," proclaimed the professor from Leiden University. It is well known that conjugated double bonds, as are those found in PAHs, conduct electricity (the best example is graphite). The flow of electric current – in the form of electrons, hydrogen, sodium, potassium, calcium, adrenaline or any other ion – through cell membrane is the basis of all energetic processes in the cell. It doesn't seem very unlikely that IOM spherules did evolve into something alive after all.

But what is "alive" and what is life anyway? Organic matter, to continue our story, came from the void in the form of IOM and arrived on a warm and wet planet, the third one from the Sun. The process of weathering released it from its space vehicle, the meteorite, and now we can see it as a tarlike layer of dirt floating on the water. Its destiny is to end its existence on sandy beaches of our primordial ocean. But this is not what actually happened. Kerogen developed into something water-like, something that could persist on a wet planet. It adapted itself to a new environment – and more.

Organic matter gradually began to grow and reproduce itself, eventually inundating the planet with an immense number of forms. What is life? Nothing more than space kerogen's struggle for existence.

Space snow

The main problem with all the theories dealing with the emergence of life on this planet is how to solve the question of synthesis and concentration of organic material. At present there are 2.42×10^{12} tons of biomass, dry weight, on the planet, the majority (99%) of which is concentrated in green plants growing on the continents. But life didn't find its place on *terra firma* until the Silurian period (about 400 million years ago), and green plants were certainly not the first organisms. Dry mass of all the animals and microorganisms in all the oceans accounts for three billion tons (3.2×10^9 t); a remarkable amount if it had to be synthesized from volcanic fumes or collected from meteoritic debris in the form of preformed amino acids and nucleic bases.

But kerogen is quite a different story altogether. Curiously, polycyclic aromatic hydrocarbons are the most abundant organic compounds found in outer space. Many synthetic routes for their formation have been proposed. According to the high-temperature synthesis scenario, organic compounds originated in the inner Solar System prior to planetesimal accretion. Kerogen could as well have been formed on icy particles by action of solar UV radiation or synthesized on comets from formaldehyde by a kind of formose reaction, as already mentioned. This is a problem in astronomy (or astrobiology) and whatever may turn out to be the final answer, it would not change these important facts: that there is an immense quantity of PAHs in interplanetary space and that a good portion of them found their way to Earth during its early period.

They arrived with comets and meteorites, and even more had come riding on smaller cosmic bodies, if they could be called bodies at all. I'm referring to bacterium-sized (≈ 10 µm) interstellar dust particles, or flakes, which are still steadily arriving through our atmosphere. In contrast to meteorites, which contain less than 5% organic material, space dust is nearly 50% organic, mostly in the form of IOM.

By this steady flow of all that "garbage" from the sky, it is estimated that each year our planet is enriched with 10^7 kg of new carbon. This space fallout used to be considerably greater during Earth's formation, up to 10^9 kg/yr. All the carbon in the oceans may have been accumulated in a just few thousands of years!

All theories of life's origin so far tried to explain the emergence of life by some kind of "pond," which was necessary to concentrate organics and to provide relatively stable conditions for their development. Curiously, for the "aromatic world" theory such a pond is not necessary. It has been mentioned before that PAH particles have the structure of a hollow sphere. Inside this sphere a concentration of low-molecular material could take place, creating a chemical system of its own, with little interaction with the surrounding ocean waters, and even less with the processes occurring in other particles.

And so life began. Or are we still too fascinated with formidable skills of our last chef?

---

660   Mill. yrs from present

Rodinia broke up

630

Mill. yrs from present

oxygen had risen to 50% of
present atmospheric level,
formation of the ozone layer

8. THE TABLE

Which theory is the right one – and what do we know about the first life on Earth at all?

Pater Halili belongs to a group of very special people who left a lasting impression on me in my youth. He was a Roman Catholic priest but I have never heard him speak of God. Pater Halili was absolved of all of his ecclesiastical duties after he had killed a man during his stay in Kosovo, a tumultuous region of Balkans. Although the judicial court absolved him of any guilt, for he did it as an act of self-defense, the court of church held a different view; no-one with the blood of man on his hands should be given permission to speak in the name of God. So what did he talk about? Not God, but food.

"I believe that paradise is a long, long table laden with all kinds of sweets," he used to proclaim following a regular dinner at my aunt's, who was – needless to say – an excellent cook (and a good Catholic). "You stroll along the table, taking this and that and you'll chat, chat, chat…" To elaborate on his eschatology, he would usually end his sermon with a sentence: "Would we have this talk if we didn't happen to join together by the table?"

Pater Halili is now in paradise, probably in his very own particular version of it. We, however, are still on this earth considering which food to choose from a (paradise-like) table heavily laden with absolutely everything. If we partook of just one dish it wouldn't be enough; if we ate all that is on offer it would be crazy and impractical, and going home hungry is simply out of the question.

This "smorgasbord" we are talking about are our theories aimed at the question of how life originated on our planet. One theory is not enough, and to accept all theories would be crazy, a logical impossibility. So, what should we do?

600

the oldest fossils of complex
multicellular organisms
(Ediacara)

This chapter is devoted to that question. We have met many chefs, tasted many dishes and now we have to decide on our meal. But how do we compose the right menu?

Theories waiting for an answer

Scientific theories are quite unlike fashionable garments. They do not get accepted simply because they happen to be in vogue. They are not like cars either. Their acceptance is not founded on being technically superior to older models. Scientific theories are a lot like life itself: they live and die, they have their fathers and mothers, sons and daughters, brothers and sisters; they can be in trouble or in full strength, sick, distorted, perverted but also healthy and bright. And they can even die for a time only to be resurrected in a new glory.

No one seriously doubts that oil deposits had been formed during the Mesozoic era, and that Mendeleev's nineteenth-century hypothesis on the inorganic origin of petroleum should be viewed merely as an old-fashioned speculation. Nonetheless, a hundred years later Thomas Gold independently arrived at the same conclusion, proposing that the first food for life, and effectively the first organic material, was methane released by chemical processes in the depths of Earth's crust. Urey and Miller assumed that organic matter originated through synthesis from gases in a primitive, Jupiter-like atmosphere. Although we now doubt that such an atmosphere ever existed on Earth, we still accept the hypothesis that volcanoes provided the gaseous ingredients necessary for Miller's synthesis. Space impactors weren't even mentioned in the original Urey-Miller theory, but their impacts might have perturbed the atmosphere to a very high degree and, accordingly, many organics could have been formed that way.* Theory of panspermia has been abandoned, but then it appeared again in the guise of prebiotic synthesis in comets, meteorites and asteroids. Attempts of nineteenth-century scientists to create life by mixing inorganic chemicals, producing grass, mushrooms, and similar "live" forms in their "chemical gardens" may now

---

*By this impact CO_2^- rich atmosphere produces CO, O_2, H_2 and NO, in contrast to CO-rich atmosphere which would yield CO_2, H_2, CH_4, HCN, NH_3, and H_2CO. It was calculated that an impactor of the size of the one which killed the dinosaurs 65 millions of years ago would have produced from 10,000 to up to hundred million tons of HCN in the early Earth's atmosphere.

	570	560
	Cambrian explosion; the first fossil evidence of comb jellies, sponges, corals and sea anemones	the earliest findings of fungi

look like child's play but, by the end of the last century, "chemical garden experiments" with iron(II) chloride solution were performed in order to support iron-sulfur-world theory.

Such a state of affairs (or minds) is not unusual in science. "A theory is only an apparatus in which a chemist performs his experiments," wrote a Croatian chemist, Professor Bubanović in one of his books on popular science at the beginning of the last century. Scientific theories are neither weird speculations nor undeniable truths. They have to be brought into accord with the results of experiments and observations, with facts. If that is not possible, a new theory has to be proposed, possibly by resurrecting one which happens to be nearly forgotten.

But there is another possibility. A theory might be too narrow to encompass all experimental findings. Two, three or even more theories should suffice, but is it possible to put all of them into accord with each other? It cannot be claimed that extraterrestrials planting the seed of life on Earth, and that life being a natural consequence of geochemical processes can both hold true at the same time. We might assume that the Universe is God's creation and that life on Earth is but an expression of divine will, but then it would make no sense at all to develop a scientific theory of the origin of life, because it would contradict our belief that life emerged as a miracle of God. This is not the way that leads to a any greater understanding of Nature.

But there is another way, the way of modesty and sincerity. What we know, we know, what we don't know, we don't know. We positively know that there is life on Earth. We also know that we have no material proof of life elsewhere in the Universe, which leads us to a supposition that it originated on Earth. We also know that life is a very complex natural phenomenon, and so the problem of its origin cannot be solved by asking simple questions. When did the first living cell appear? How were the first proteins and nucleic acids synthesized? Which one was the first enzyme? All those questions are bad questions, because they are based on unproven presumptions. We presuppose that we know what we really know not. We don't really know what the first living cell looked like, and we know even less of its chemical composition and metabolism. We know next to nothing about the first half billion years of Earth's history during which life emerged. The same time span divides us from the Silurian period, when the continents were still desolate and barren of all life. How could we, living on land in its present form, possibly have any knowledge of trilobites and the like monsters from the past? All organisms evolved from a common ancestor, usually abbreviated as LUCA (last universal common ancestor), as is evident from the fact that all inhabitants of this planet share essentially the same biochemistry. LUCA is just a name, however. What do we know about it? Nobody can positively say what LUCA looked like. And what was there before the LUCA?

But something may still be said though. There is a lot of experimental data, there are many hypotheses and theories, many experiments performed and many results published. One experiment is not enough, one paleontological finding can't be decisive, and no single theory could possibly be entirely convincing. But a bunch of them put together just might be. Is it possible to unite all of them in one coherent picture which could provide a convincing answer to the question of the emergence of life on this planet?

Let's try!

In the beginning...

The history of Earth may be told in just six sentences. Earth originated by accretion of asteroids and planetesimals 4.6 billion years ago, and life on it emerged half a billion years later, as the fallout from space died down. Its atmosphere was originally free of oxygen, the gas which then accumulated with the appearance of blue-green algae about 2.7 billion years ago. This free oxygen in the atmosphere enabled development of multi-cellular organisms, leading to the "Cambrian explosion" of life forms 570 million years ago. The first plant life appeared on land during the Silurian period (395 – 435 million years ago), first land vertebrates, amphibians, showed up in Devonian (345 – 395 mill. yrs. ago) followed by first reptiles which evolved further into dinosaurs who reached their pinnacle in the Mesozoic. Their comfortable existence came to an end with the fall of an asteroid near Yucatan peninsula 65.5 mill. yrs. ago. Their extinction ushered the age of mammals, which include our species, *Homo sapiens*, whose traces may be found up to 200,000 years in the past.

But this short history of life isn't by any means complete or even particularly accurate. Life's story on this planet is quite dramatic indeed; there were numerous periods when it found itself very close to extinction due to cataclysmic volcanic eruptions, asteroid impacts, sudden climate changes

521
510

the oldest
fossils of
trilobites

A short overview of Earth's history

Aeon	Era	Period	Million years ago	State of the planet	State of life
Phanerozoic	Cenozoic	Quaternary	0 – 1.6	Ice age	The first man, formation of present biosphere
		Tertiary	2 – 65	Earth surface similar to its present appearance	Development of mammals, first cereals
	Mesozoic	Cretaceous	65 – 135	Formation of modern mountain ranges	First angiosperms Extinction of dinosaurs
		Jurassic	135 – 190	Mild climate	First birds
		Triassic	190 – 225	Regression of the sea	First mammals and bony fishes
	Paleozoic	Permian	225 – 280	Intensive tectonic activity and climate changes	Extinction of trilobites First ammonites
		Carboniferous	280 – 345		First reptiles
		Devonian	345 – 395		First amphibians
		Silurian	395 – 435		First fish and land plants
		Ordovician	435 – 500		First echinoderms and vertebrates
		Cambrian	500 – 570		First shelled fossils

490   480

the oldest fossils of vertebrates with true bones (jawless fishes)

Aeon	Era	Period	Mill. yrs ago	State of the planet	State of life
Precambrian	Proterozoic	Late	570 – 900	Snowball Earth extreme ice age period	
		Middle	900 – 1600	Warm climate	First multicellular organisms
		Early	1600 – 2500	Increase of atmospheric oxygen	
	Archaean	Late	2500 – 3000	Ice age	First cyanobacteria
		Middle	3000 – 3400	Warm climate	First photosynthetic organisms
		Early	3400 – 3800	Cessation of heavy meteoritic bombardment	First living forms
	Hadean	Not divided	3800 – 4600	Liquid water Continental crust (?)	Origins and successive extinctions of life (?)

(geological past is brimming with them!), floods, bursts of cosmic radiation, and many other known and unknown catastrophic events. After listening to all the horror stories from the planet's past, it seems almost a miracle that life still exists and, what is more, that it had managed to develop into those highly complex forms which now shape our modern biosphere.

What is most surprising in the timeline of life is its uneven span. To someone without proper education in geology the figure of 4.5 billion years has the same meaning as 4.5 billion dollars would have to someone without a comparable amount of money in their bank account; both figures are beyond their imagination. This money would buy three bicycles for every minute of one's lifetime, as French comedian Fernandel took pains to calculate for another huge sum in one of his movies, and 4.5 billion years would contain

450

438

Andean-Saharan glaciation

1st mass extinction (Ordovician/Silurian)

fifty million of our human life spans. To familiarize oneself with geological time scale is the most difficult step in studying geology, which is why some kind of metaphor or the other is usually found in geology books intended for layman readers.

If we take one second to represent a million years, then our Earth would be only 4,500 seconds or 75 minutes old. In this case, life in its developed, Cambrian and post-Cambrian forms, would have existed for only 10 minutes, dinosaurs went extinct a minute ago, and our species, *Homo sapiens*, is less than 1/5 of a second old! The period covered by our civilization, from ancient Sumerians and Egyptians till now, is not even perceptible – it lasted 10,000 years, or just 1/100 of a second in our 75-minute geological history!

But let us take this further. We hardly know anything about what happened in the first eight minutes of Earth's history. And for the next three minutes, when life finally came into being, we have no paleontological evidence at all. First such evidence is found in the isotopic composition of the 3.8 billion years old stromatolites, and the 3.5 – 3.3 billion years old (17 minutes after the formation of Earth, on our time scale) microbial mat found in Josefsdal Valley, near the asbestos mining village of Msauli in the Barberton Greenstone Belt, South Africa. Prior to this discovery, the oldest fossils known were those of *Eobacterium isolatum*, a bacterium 0.56 × 0.24 µm in size, found in black shales of the Fig-Tree formation in Central Transvaal, also in South Africa. They are 3.1 billion years old.

"The mat-like structure was formed by multiple layers of parallel filaments that have a constant diameter of 0.25 µm along their lengths and that are thickly coated with a heterogeneously textured film that ranges from ropy, granular, smooth to holey," is the description of 3.5 billion year old fossils from Josefsdal Valley. "The filaments are commonly interwoven with each other, producing thicker fibrous structures up to several micrometers in diameter." Also found near the filament was a small group of rod and vibroid-shaped structures, 1 µm in diameter and 2 – 3.8 µm in length, very similar to modern bacteria (bacilli).

Those filaments were quite flexible, as indicated by their being curled around embedded volcanic and other mineral particles. This indicates that they sedimented in shallow water, on a beach in a lagoon at the foot of a volcano. The volcano eventually erupted, and its ash and lava covered the organisms, conserving them for all eternity.

434	420	407
the first primitive plants move onto land	the oldest fossils of sharks	the oldest fossil insect (*Rhyniognatha hirsti*)

Research did not end with these observations however. Analysis of filaments uncovered carbon in the form of kerogen (similar to that found in meteorites) and its isotope composition clearly showed that these organisms engaged in photosynthesis – they used solar radiation to produce organics. This supposition was further confirmed by the orientation of filaments, which were arranged vertically, in the manner of modern photosynthetic filamentous microorganisms. But how could they be photosynthetic, if there was no free oxygen – that is O_2 – in the atmosphere at that time? And how could they use solar energy when the first blue-green algae, as mentioned earlier, appeared just 2.7 billion years from the present, nearly a billion years later?

It is futile to attempt finding traces of photosynthetic enzymes in the spectrum of kerogen which is three billion years old. However, vibroid-shaped structures are commonly attributed to fossils of sulfur-reducing bacteria, those thriving on sulfates. Yet even older microfossils were found this year (2017) in hydrothermal vent precipitates at Nuvvuagittuq supracrustal belt (NSB), Quebec, Canada. Their age was estimated to be at least 3.77 billion years, and they are similar by their form to photoferrotrophic bacteria, the ones that use light to convert Fe^{2+} into Fe^{3+} ($Fe^{2+} + hv \rightarrow Fe^{3+} + e^-$). And from these findings something could still be discerned – or at least supposed.

Photosynthesis without oxygen

One of my father's favorite books can still be found on my bookshelf. Its title is *Leksikon prehrane* (Lexicon of Nutrition) by Ernst Mayerhofer, a professor at Zagreb School of Medicine. The book was originally written in German but as far I know it was never published in that language. But that is not the only curious thing about the book. Its author was an Austrian pediatrician and a lecturer at the University of Vienna who came to Zagreb to help build a children's hospital there in 1922. The lexicon was published in the war year of 1944 and, despite its peaceful title, in many ways it is a kind of wartime book. The reason for this is because it is not a book about delicious food, good cooking, skilled service, exotic cuisine and the like, and neither is it a book about healthy food, low fat and low sugar diets. It is a book about hunger – how to provide the best energy and protein intake from very scarce resources. It is also a book about *Ersatz*, German for substitute. The first substitution is, or course, replacing human milk with cow milk in

400	395	390
the oldest fossils of ammonites	the first findings of lichens, stoneworts	

the feeding of babies. But there are other, more exotic substitutes such as hay for flour in bread baking, sulfuric acid for vinegar (acetic acid) for preserving cucumbers, and marl for butter in bread and butter sandwiches. Hunger asks no questions, as a Croatian proverb goes. There is also an article about brewing of beer during the Great War from anything on which yeast can grow, including potatoes, beans and horse-beans, carrots and chestnuts. Of course, Professor Mayerhofer did not recommend the use of all those substitutes, he merely recorded them. But substitution is not only a natural consequence of food shortages; it is also one of the great principles of nature.

Life evolves by adaptation of species to new habitats. They seek new opportunities or, in other words, they try to adapt themselves to a new, at first very hostile environment. Whether the word "adaptation" is understood in the former or the latter sense, the key terms remain the same: substitute and substitution. Semi-aquatic animals evolved from their land ancestors when they substituted sea weeds for land grass, carnivores turned into carrion-eating scavengers and *vice versa* depending on available food. Primates, an order of animals we belong to, evolved from insectivora – animals feeding on insects, slimes and worms – into omnivora, those that may sustain themselves on all kinds of food. Carnivorous plants substitute minerals found in bodies of captured insects for minerals they used to extract from earth, which enables them to grow on poor soils. Blue-green algae or cyanobacteria may grow in the dark, as saprophytes, if adequate nutrients (e.g. polluted water) are provided.

The same natural law holds true in the realm of biochemistry. Some animals require vitamin C while others do not. Those that do require it, like humans, must eat fruits and vegetables, and those which do not, like mice, may thrive on foods wholly deprived of it. There is also a set of essential, as well as a set of non-essential, amino acids. The first set encompasses amino acids abundant in common food, such as phenylalanine or threonine, for which there is no need to synthesize them in body. Others, the non-essential amino acids, are synthesized from keto acids, and they in turn from carbohydrates, thus substituting carbohydrates for scarce amino acids. But just as in the case of carnivora and scavengers, there is no strict distinction between the two groups, animals and amino acids. A rat can synthesize arginine by itself, but this synthesis is not sufficiently productive during the rapid stage of a rat's development; therefore for that particular animal arginine is both an essential and a non-essential amino acid.

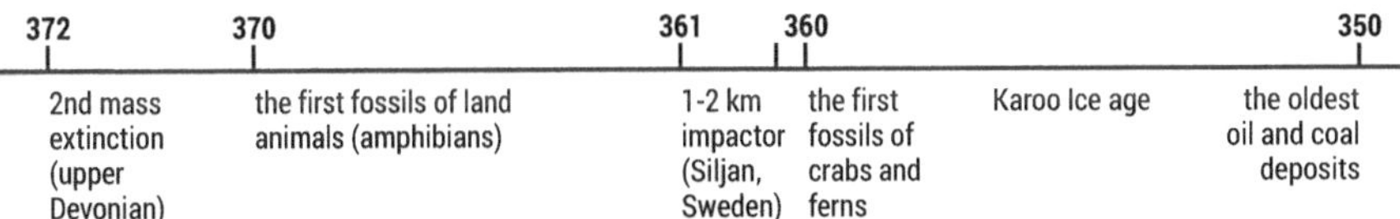

This natural process of substitution led Oparin, as described in the third chapter, to propose the development of "metabolism" of his droplets by successive depletion of nutrients: in transformation of A into B, A was replaced with C after A had been depleted and the route $C \rightarrow A$ was established. This course of events was also accepted by the followers of chemoautotrophic theory, though involving different chemicals and different reactions.

The first source of energy was, according to chemoautotrophic theory (see chapter six), the reaction of iron(II) sulfide, FeS, with hydrogen sulfide, H_2S:

$$FeS + H_2S \rightarrow FeS_2 + 2H^+ + 2e^-.$$

The released electrons (e^-) provided the driving power for all the other reactions in the pioneer organism. These electrons may be used for conversion of carbon dioxide into organic material, here represented simply by formula CH_2O (not to be confused with formaldehyde!):

$$2FeS + 2H_2S + CO_2 \rightarrow 2FeS_2 + CH_2O + H_2O.$$

But what do we do when iron(II) is depleted?

In order to answer this question we don't have to travel four billion years into the past. The hydrothermal system of Yellowstone Park houses bacteria which grow on hydrogen sulfide while also using light (hv) as a source of energy. In their habitat, "sulphuretum," they convert hydrogen sulfide into sulfuric acid:

$$H_2S + 2CO_2 + 2H_2O + hv \rightarrow H_2SO_4 + 2CH_2O.$$

What does this mean? Firstly, iron(II) is replaced with a photon (light), and as a result electrons of sulfur are better utilized. Namely, in the "iron-sulfur-world reaction" one molecule of hydrogen sulfide produces two electrons, but in the second, photosynthetic reaction eight H2S electrons are available for synthesis. Effectively, in order to convert one molecule of carbon dioxide into organics, two molecules of hydrogen sulfide are required for chemoautotrophic reaction, but in phototrophic conversion from one molecule of hydrogen sulfide, two molecules of fixed carbon

---

338

the oldest fossil
of reptiles
(*Westlothiana lizziae*)

330

diversification
of amphibians

rise of CO_2 in atmosphere

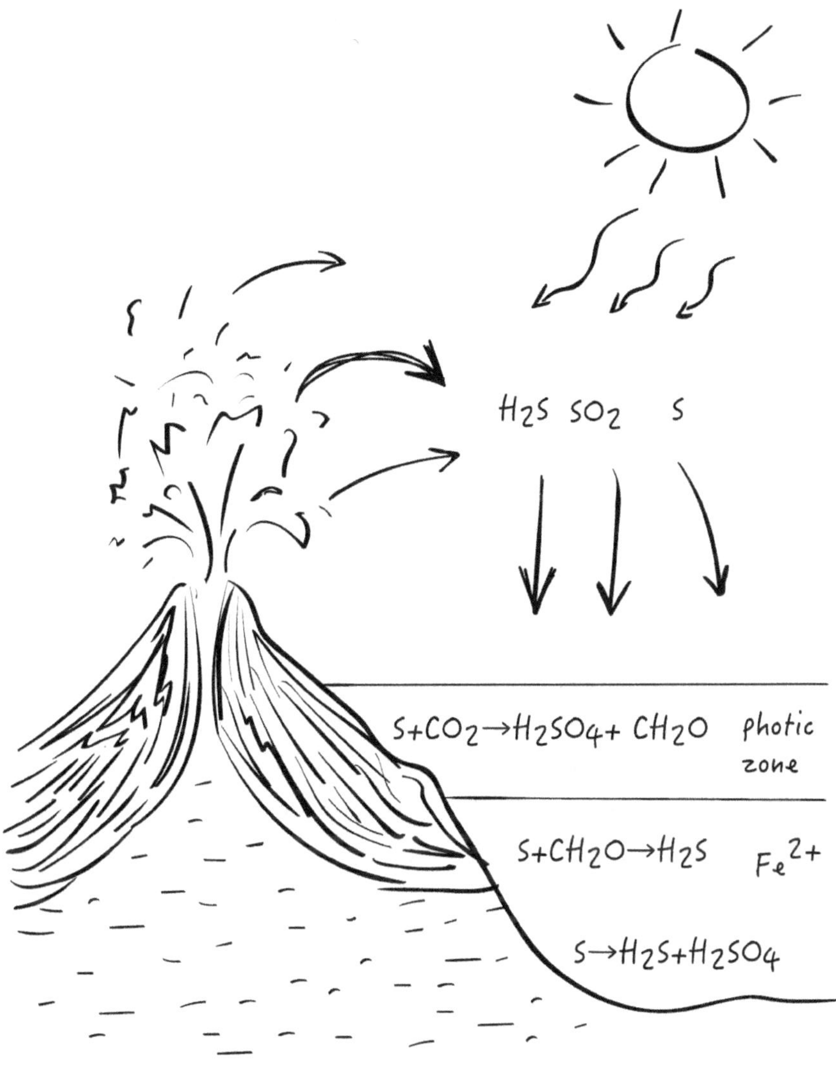

Life very probably evolved on the slope of an active volcano, which provided sulfur compounds, H_2S and SO_2, as well as elemental sulfur, S, needed for photosynthesis by photoautotrophic organisms. Organic matter, marked here as CH_2O, provided food for heterotrophic organisms thriving below the photic zone (adopted after D. E. Canfield et al., *Phil. Trans. Roy. Soc. B* **361** (2006) 1819–1836).

300

formation of super-continent Pangea | the first fossils of spiders | evolution of beetles are produced. There is also no more need for FeS, and so the "nutritional requirements" of the first organisms were radically reduced. This is a much better recipe than any found in the hunger-inspired lexicon of Professor Mayerhofer!

But there is another possibility, more appropriate for the early Earth.[*] Namely, it is not very probable that primordial ocean contained much free hydrogen sulfide for it would be immediately precipitated as FeS. But there was certainly plenty of elemental sulfur, because it was obtained by the reduction of volcanic sulfur dioxide. Thus it is possible to write an equation quite similar to the previous one:

$$2S + 3CO_2 + 5H_2O + h\nu \rightarrow 2H_2SO_4 + 3CH_2O.$$

But what was it that was really happening in this presumably the first of photosynthetic, however anoxygenic cycles? This is made clear once they are compared with the oxygenic one (photosynthetic reaction of green plants):

$$CO_2 + H_2O + h\nu \rightarrow CH_2O + O_2.$$

Again, the diet was reduced further. Now only carbon dioxide and water are required. But what is really going on here, strictly chemically speaking?

In all of the above mentioned biosynthetic pathways electrons are transferred to a CO_2 molecule thereby reducing it to CH_2O ($CO_2 + 4e^- + 4H^+ \rightarrow CH_2O + H_2O$), the only difference being where the electrons come from. In the first two photosynthetic reactions they come from sulfur, either elemental or in the form of hydrogen sulfide; in the third reaction they come from oxygen, or to be more precise, from oxygen in a water molecule ($2H_2O \rightarrow 4e^- + 4H^+ + O^2$). In effect, the iron-sulfur-world is firstly reduced to a sulfur-world and then to a world without any sulfur at all.

*Many alternative ecosystems were proposed as well, such as the hydrogen-based one, in which electron donor is elemental hydrogen (H_2), and a nitrogen-based ecosystem, which was driven by photolytic reduction of nitrogen (N_2) into ammonia (NH_3). But I must admit I don't see much benefit in such speculations about life on our planet (for Jupiter it's a different story), because they have no support in iron-sulfur-world theory. It is not enough that a phototrophic metabolism simply may exist, but it is necessary that it is conceivably capable of evolving from a more basic metabolic pathway.

But there is yet another "reductive diet" for pioneer organisms. It is what gave form to the iron world.

There were many sulfur compounds on the surface of primitive Earth, and they mostly came from volcanic eruptions, but there were even more iron(II) salts, which were released from minerals. They both formed a primordial iron-sulfur world. After a while this volcanic activity ceased and with it the inflow of new sulfur ceased as well. Protocells began to starve. And the remedy for starvation is substitution: they replaced sulfur with light.

This substitution resulted in the following phototrophic metabolism:

$$4Fe^{2+} + CO_2 + 7H_2O + h\nu \rightarrow 4FeO(OH) + CH_2O + 8H^+.$$

What is happening here? The electrons aren't coming from either sulfur or oxygen, but rather from iron(II) ions. In this process iron(II) is converted to iron(III), that is $FeO(OH)$, a mineral called goethite. And as is the case with previous hypotheses, this one was also inspired by the discovery of actual bacteria with such metabolism – and supported by finding of 3.77 billion years old fossils in Canada.

This system is notoriously inefficient though; these bacteria had to "eat" four ions of iron to fix only one molecule of carbon dioxide, but even a reductive diet such as this was sufficient enough to allow diffusion of life into the ocean, far away from the fuming volcano. And there was plenty of food in the ocean: it is estimated that Archaean concentration of iron(II) ions was in the range of 40 – 120 μmol/L, that is about 2 – 7 g/m^3.

Nice theory you got there, however...

Scientists have made many efforts to propose and elaborate on the first photosynthetic metabolisms and ecosystems based on them. They usually imagine a lagoon near a volcano (like the one in South Africa 3.5 billion years ago, where the oldest fossils were found), and a mountain slope whose weathering would provide the necessary nutrients. Hydrogen sulfide is provided by hydrothermal vents on the bottom of the lagoon as well as by hydrothermal springs near its photic zone, its surface that is. Sulfur came from the air and blended with surface waters. But one major problem still remains unresolved.

252	250		245	240	
3rd mass extinction (the largest) (Permian/Triassic)	evolution of flies		the earliest fossils of ichthyosaurs	break-up of Pangea	

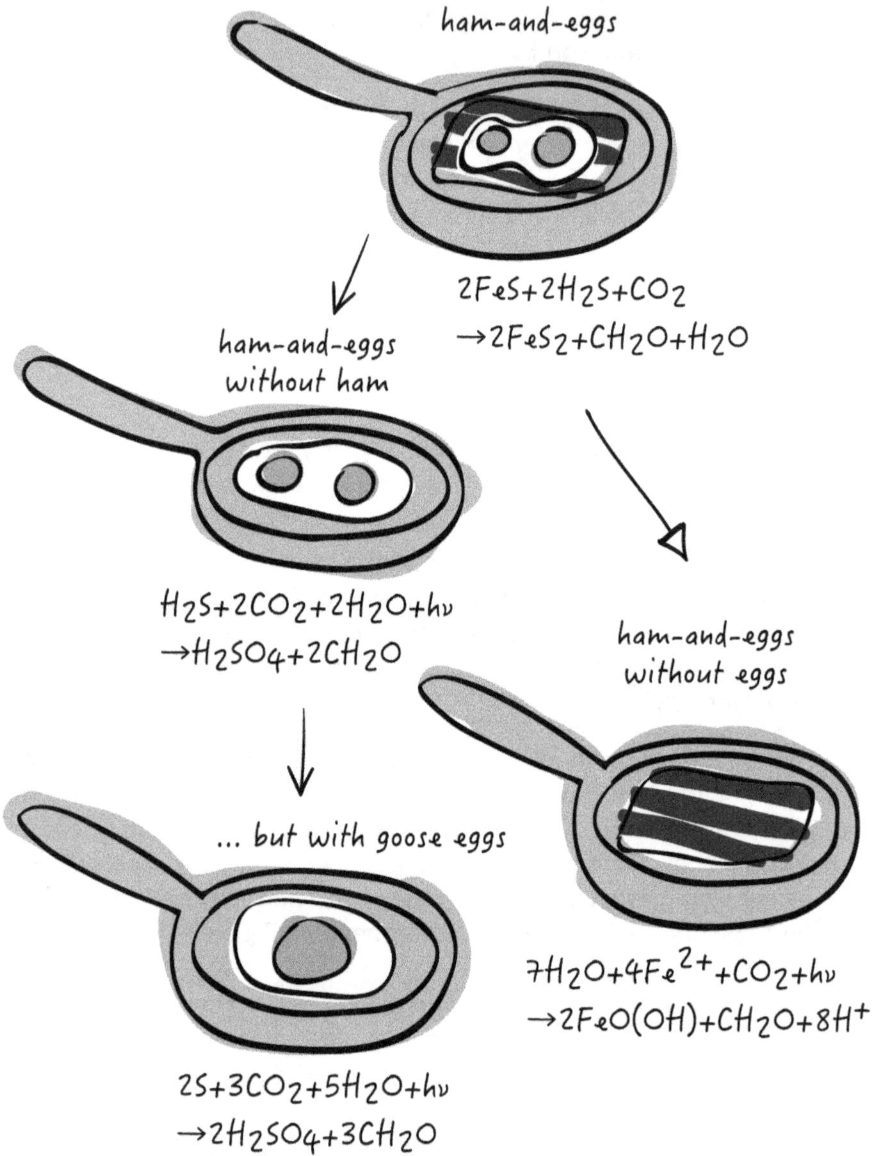

Evolution of a photoautotrophic metabolism using ham-and-eggs simile. If you don't have any eggs, you can fry only ham, and if it's ham you're missing then you're stuck with plain eggs. If you wish to improve your ham-deficient diet you may use larger and very fatty goose eggs.

225	220	210	208
findings of the first mammal fossils	the oldest fossils of turtles	the oldest dinosaurs fossils	4th mass extinction (Triassic/Jurassic)

How much of the hydrogen sulfide and elemental sulfur (i.e. sulfur dioxide) was available to feed the first phoautotrophic organisms?

Volcanic emissions are very rich in sulfur compounds, and so this presents us with a very interesting sulfur chemistry such as the one we might expect to find in hell. Roman scholar Pliny the Elder, author of *Historia naturalis* (Natural History), died at the age of 56 from "volcanic plumes" while attempting to explore the crater of Mt Vesuvius which happened to be erupting at the time (as it was busy burying the towns of Pompeii and Herculaneum). It is not clear if he perished from hydrogen sulfide or from the less toxic, though more irritant, sulfur dioxide. Both gases are present in volcanic emanations, but their ratio varies depending on the particular volcano, or more precisely, on its position on the tectonic plate and mineral composition of the rock bed it rests on. Fumes emitted by Mt Etna are composed of 25.2% SO_2 and 3.4% H_2S, as well as a small fraction of free hydrogen, H_2 (0.5%). Fumes of the African volcano Nyiragongo are very rich in hydrogen (14.9%), with moderate levels of sulfur dioxide (18.3%) and hydrogen sulfide (1.6%). Additionally, there are volcanoes with less than one percent of sulfur compounds in their fumes, like 0.08% SO_2 and 0.14% H_2S for Tolbachik on Kamchatka peninsula. With such great differences in their compositions it seems that it wouldn't be prudent to perform extensive statistical calculations on them; but this was already done by other researchers and I will duly report their findings here. The average is 3.2 for volume ratio of $SO_2:H_2$ and 5.5 for $SO_2:H_2S$. Each year volcanoes produce 13 – 20 million tons of sulfur dioxide, and 4 – 35 million tons of hydrogen sulfide, the latter mostly coming from subaqueous volcanoes. (Compare this with 120 million tons of anthropogenic annual emission of sulfur dioxide.)

These quantities are quite trivial, on a geological scale that is. The primary production of modern oceans is about 50 billion tons of fixed carbon a year. It was calculated that even if volcanic outgassing was ten times the present rate, the production of sulfur-based ecosystems would still be a thousand times lower than that of oxygen-based marine ecosystems that is of those based on the production and utilization of free oxygen.

Much better results were obtained for iron-based ecosystems however, $1.7 – 5 \times 10^{14}$ mol/yr or $2 – 6 \times 10^9$ t/yr of fixed carbon, reaching up to 10% production of present marine ecosystems. But in spite of this, there is no doubt that for the first two billion years of its history Earth used to be a rather desolate place.

diversification of dinosaurs   evolution of true bony fishes

Opportunistic life

The story of the cloning of animals, and eventually human beings, did not begin in June 1996 with the birth of a female domestic sheep named Dolly, but in September 1973 with a phone call to David M. Rorvik, a notable science author, at his cabin on Flathead Lake in the mountains of Western Montana. The caller was a childless multi-millionaire, who made a fantastic proposal: Rorvik was to recruit a medical team to create a clone of himself, the millionaire client. Such a proposal necessarily opens many questions, of scientific as well as ethical nature. Rorvik then wrote a book about the event, *In his Image – the Cloning of a Man*, which became a bestseller and was advertised with blurbs such as "Humanity will never be the same again," or "A turning-point in the history of the species." It doesn't seem at all impossible to duplicate a human being from a single cell extracted from his or her body, and certainly not after Dolly the sheep came to be. In fact, it seemed so probable that many people still believe that Rorvik's book is a genuine report, rather than a clever science fiction spoof.

Cloning is not that complex a technique. In its essence, it is nuclear transfer, transplanting one cell's nucleus into another cell. The first such "transplantation" was performed in the 1960s, when an unfertilized frog's egg was subjected to UV treatment in order to destroy its nucleus and into which another nucleus from a tadpole intestine's epithelial cell was transferred. A tadpole developed from that egg and eventually grew into an adult frog – which was the exact copy of the frog whose nucleus had been transferred.

But this isn't entirely true nor accurate. In reality there are no exact copies of frogs, sheep, humans or whatever. Experiments with frogs were performed in order to prove that all hereditary material, that is genes or DNA, is located in the cell nucleus only. That may be true, but it is not absolutely true. One portion of cell DNA is located in its mitochondria, in the form of famous "mitochondrial DNA," a term often encountered in TV dramas such as "Exhibit A" or "The Body of Evidence." This mitochondrial DNA also provided the proof that once there had existed a small population of "Eves" from which all modern humans can trace their descent. It was possible to make this discovery because during conception, that is fusion of male and female gametes, it is only the cell nucleus which is transferred from spermatozoa to oocyte. Offsprings of such a union therefore inherit mitochondria

evolution of moths and wasps

fossils of
archeopteryx

from their mother only, and consequently they share her mitochondrial DNA. The same happens in nuclear transfer; the cloned and the donor cells have the same DNA in their nuclei, but not the same DNA in their mitochondria. This creates much confusion because the new nucleus cannot communicate well with the original mitochondria, because mitochondria also need nucleus DNA for their proper function.

Mitochondria serve as power plants in all aerobic organisms except prokaryotes. They are the site of citric acid or tricarboxylic acid (TCA) cycle which performs the final conversion of ingested food into carbon dioxide and water. Mitochondria resemble cells in many aspects, for they have membranes as well as genes in the form of DNA. This led to a hypothesis, fully accepted today, that mitochondria developed from the first aerobic cells which found their last refuge in the bodies of their anaerobic hosts. And they made a clever deal: aerobic cells will protect the anaerobic ones from murderous effects of free oxygen, and the anaerobic party will in turn provide the aerobic one with all nutritional requirements it requires. This deal then turned even more in favor of anaerobic cells because mitochondria obtain much more energy from the same amount of food. It is easy to calculate that glucose releases ten times more energy through conversion into carbon dioxide than into lactic acid.

This business-like relationship between organisms is known as symbiosis. Symbionts are mutually dependent and they both gain some advantage from their relationship. Symbiosis is not, of course, limited to mitochondria and their host cells only. Chloroplasts and other plastids were symbionts and then descendants of organisms which went extinct hundreds of millions of years ago. "Struggle for existence" and "survival of the fittest" are not the only laws of evolution, or – to be more precise – they shouldn't be always taken literally. Struggle for existence and survival of the fittest do not necessarily imply a war between individuals and species, *bellum omnia in omnes*. They may refer to cooperation as well: "the fittest" are quite frequently not the most aggressive or persistent, usually they are simply the craftiest opportunists.

If symbiosis is the law of life, then theories of its origin have to be symbiotic as well. This means the one theory does not necessarily disqualify the other. The theory of the development from cosmic kerogen is very convincing in describing formation of cell membranes and ultimately the cell,

<table>
<tr><td>130</td><td>120</td><td>110</td></tr>
<tr><td>the rise of angiospermous plants</td><td></td><td>fossils of the first true birds</td></tr>
</table>

but if one wants to explain the development of autocatalytic cycles, the best theory to choose is the iron-sulfur-world one. An ocean filled with low-molecular compounds, fatty and amino acids, simple sugars and the like wouldn't be sufficient to explain the beginning of life on Earth. On the other hand, production of organics by atmospheric processes in a Miller-like manner might have provided crucial aid at the right time. It is quite possible that many "live" systems did exist for some time before they were united as symbionts in the first living cell, or LUCA if you prefer this term.

But in discussing theories on the emergence of life on this planet, as in discussing any theory, one has to bear in mind Occam's razor, the much cited but vaguely understood principle postulated by English scholar William of Ockham (Occam in Latin spelling) who lived in the late 13th century. *Entia praeter necessitatem non esse multiplicanda* or "entities should not needlessly be multiplied" is how it was expressed originally and is still liable to different interpretations. Einstein said that scientific theory had to be as simple as possible, but not simpler than that. Some kind of Occam's razor is used unconsciously by every physician who is trying to find the cause of a patient's ailments in a common rather than an exotic illness. To me personally, Occam's razor is an integral part of my working routine; I try to solve each problem first with one or a two-parameter function; if that doesn't work I have to "multiply entities," that is introduce additional parameters or variables.

When we apply Occam's razor to theories presented in this book, it becomes apparent that RNA-world theory implies too much *entia* to be workable. It starts from only one entity, the RNA molecule, but this molecule is composed of at least three very different components, nucleic bases, sugar (ribose) and phosphate moiety, all of which had to be somehow arranged in a manner that seems to be quite mysterious in retrospect.

Conversely, Oparin's theory requires just one entity, coacervate droplet. Unfortunately, this entity is not sufficient to explain emergence of complex life, and the theory may be considered as simpler than what is possible, as Einstein would probably say. It is not clear how our complex machinery of life might have evolved from simple coacervate droplets, not to mention the complex catalytic and autocatalytic systems they contain, even though many "random" peptides do exhibit some kind of catalytic activity. And even worse, the simplicity of his theory cannot be corrected by introduction of some new "entities."

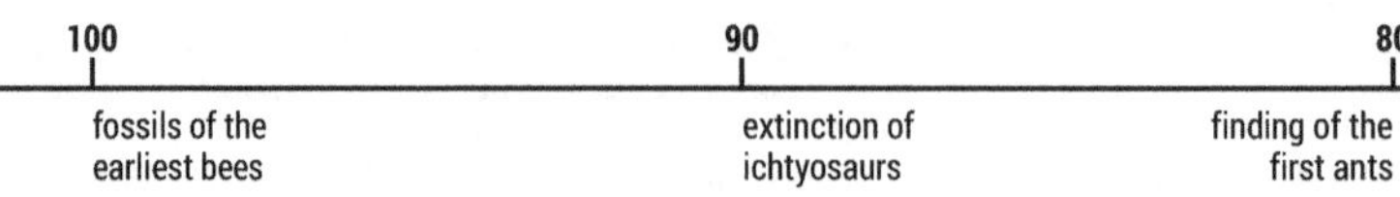

But things are different with iron-sulfur-world theory. It starts with just one substance, FeS, and only one reaction, its conversion into pyrite. In spite, or because of, this apparent simplicity, it opens a plethora of possibilities for multiplication of "entities," if necessary. Iron sulfide may be combined with catalytic sulfides of nickel, tungsten and molybdenum as well as with their carbonates and oxides. Even though iron-sulfur-world theory starts with just one reaction, oxidative formation of pyrite, it immediately introduces many others. The theory is well supported by our knowledge of metabolism found in modern organisms, as well as by geological evidence, and it readily explains the evolution of chemoautotrophic into a photoautotrophic metabolism, as already presented in this chapter. It is also based on the quite familiar process of heterogeneous or contact catalysis, widely used in chemical industry. But it does have certain flaws as well. It presupposes that adsorption of organics on the surface of sulfide particles served to enhance their catalytic activity. "Carefully chosen ligands may drastically increase the activity of transition metal catalysts by factor of up to 10^3," Wächtershäuser claimed in favor of his theory. But let us be careful here for the most important word in this sentence is "carefully." If ligands are not *carefully* chosen, they have exactly the opposite effect. "Poisoning" of industrial catalysts, as well as other contact catalysts, is a well-known phenomenon. If your car has a catalytic exhaust then you cannot use leaded gasoline because you'd risk catalyst poisoning.

The second flaw in this theory is revealed by detailed study of iron-sulfur proteins, which were seen as leftovers from the first catalytic processes. Iron-sulfur clusters which form the core of such proteins are not formed spontaneously, by simply mixing iron and sulfide ions with their respective protein (apoprotein), as is done in simple test tube experiments. In the 1990s it became evident that assembling of iron-sulfur proteins is a highly complex catalyzed process. As far as we know, assembling a bacterial Fe/S protein requires 15 components, and for a eukaryotic cell we'd need as much as 25 of them. And you can't even get simple sulfur directly from sulfide, but rather it has to be produced through a complex enzymatic degradation of amino acid cysteine. However, this may not be as great of an obstacle as it seems at first. The old sexist trope that driving cars is a man's business arose from the simple fact that first motorists also had to be mechanics and of considerable physical strength. Life expectancy of a car tire was 10,000 kilometers on average, which meant that a car was anticipated to experience a tire rupture every 2,500 kilometers. Motors were also in

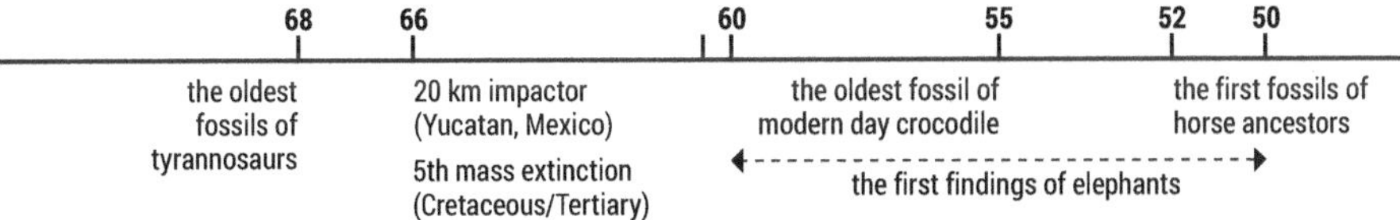

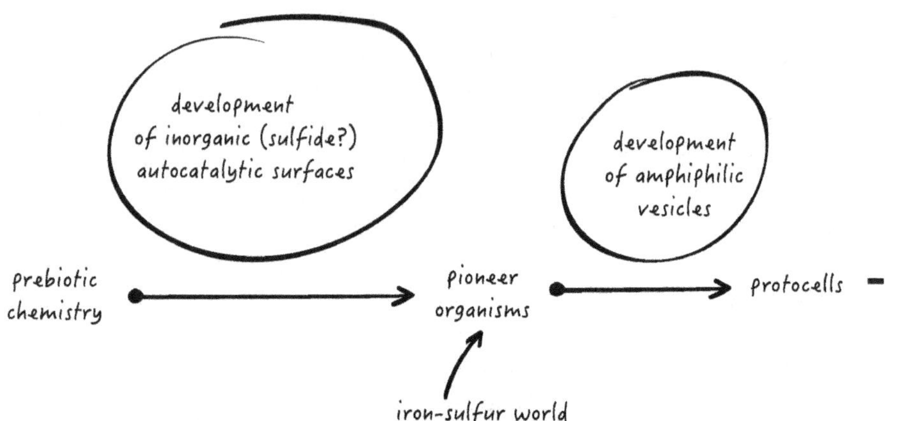

Stages in the origin of life, "a self-sustaining system capable of Darwinian evolution," as defined by NASA, or formation of the first living being(s) which may also be defined as "any autonomous system with open-ended evolutionary capacities"

need of frequent repairs, so much so that an old technical magazine I found presented an idea to provide the car with two motors, the second one to be used when first one breaks down. The same line of reasoning should be applied to first metabolic reactions, irrespective of whether they were based on iron-sulfur clusters or particles, RNA-like molecules or anything of the kind. In the beginning they had to be crude and non-selective, but they did get perfected later.

It is natural to assume an evolutionary pathway such as this. What is aerobic metabolism but simply the anaerobic one perfected, an upgrade? In anaerobic organisms pyruvic acid (or pyruvate) is converted into lactic acid, ethanol, acetic and formic acid, acetone and the like; and in aerobic ones it is converted into acetyl-coenzyme A and "burned" in the tricarboxylic acid cycle. Alternatively, anaerobic organisms evolved into aerobic ones through symbiosis with ancestors of mitochondria. And the same had to hold true for the first living cells, if they may be called cells at all.

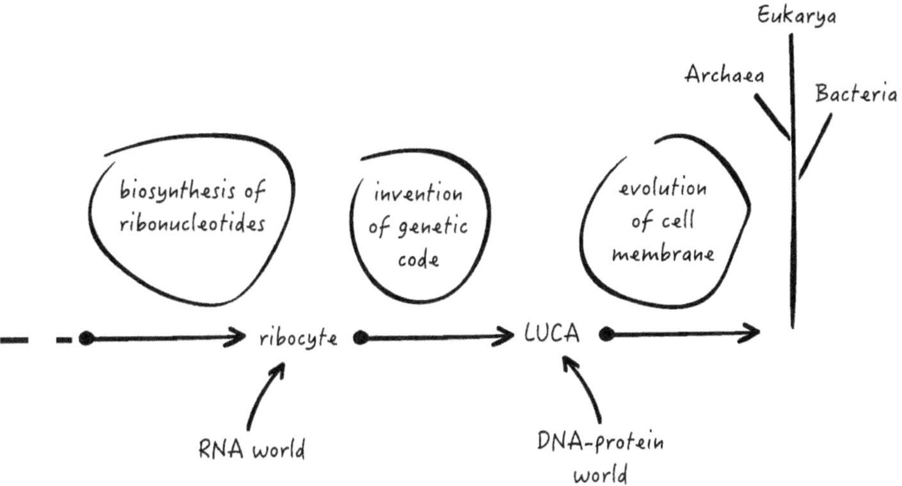

(J. Peretó). All life on Earth, as we know it, is based on the central dogma of molecular genetics, DNA→RNA→protein. But what was going on before that happened? (Adapted from J. Peretó, *Internat. Microbiol.* **8** (2005) 23–31.)

If we sum up all that we know about the origin of life on our planet, we won't find anything miraculous or unnatural there. A lot is simply unknown to us, as is the case with many fields of science. The problem is primarily psychological. We try to reconstruct events at the beginning of Earth history in test tube experiments which last but a few hours, or a few days at most, but the experiments Nature was performing in her "warm little ponds" took place over millions of years at least. How many such experiments Mother Nature performed is impossible to calculate, or maybe even imagine. Our theories propose only general principles for those experiments, like the recipe for *pašticada*, as her mother-in-law taught Marija at the beginning of this book.

Life does not have a human dimension but a planetary one. And furthermore, in its essence it is truly a cosmic phenomenon.

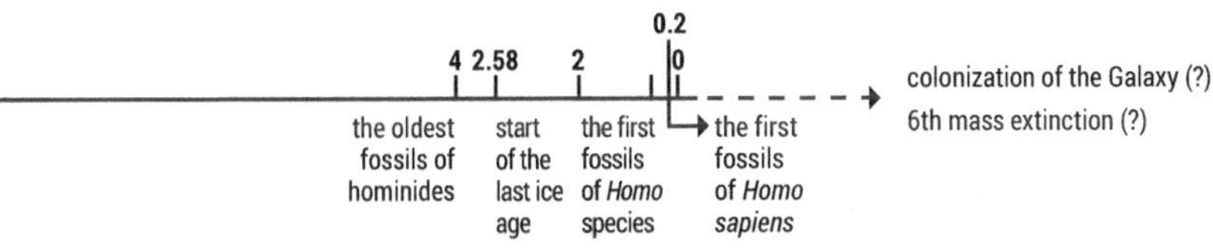

What kind of life is to be expected on other planets?

9. WEIRD OR WHAT?

Is it possible to find life on other planets? Maybe we can find traces of early Earth life on other worlds?

We've arrived at the end of a book about the origin of life, and yet we still haven't found a suitable answer to the very first question: What is life? In one respect, biology is quite unlike physics. For example; concepts such as cell, animal, plant and ultimately life cannot be defined in the same way as those of force, energy or electric charge. As a concept, Life is as vague as Beauty, Love or God. This leads us to the following conclusion: if something is as vaguely defined as life, then almost anything could be considered as alive and it would be quite possible that life exists in forms very different from anything known to us as such. Life on Earth is based on water, proteins and nucleic acids, in essence water and carbon compounds. But it doesn't have to be so, as Carl Sagan pointed out in his book *The Cosmic Connection*, first published in 1973. It is nothing more than an illusion that life must be Earthlike, he said, and this notion is just a product of our peculiar mental blindness.

There are many kinds of such blindness however, and its victims are identified by labels such as ultraviolet light chauvinists, temperature chauvinists and planetary chauvinists. Those who believe that life cannot exist without water or carbon compounds are, naturally, carbon or water chauvinists. But why should life be based exclusively on carbon and water? Carbon could be replaced with silicon, and even with germanium, and ammonia or melted sulfur may serve as good substitutes for water. Many years later David Toomey wrote the book *Weird Life* to discuss all those possibilities. But I must confess I don't really enjoy such speculations. First of all, they are in conflict with the principle of Occam's razor: why invent new entities if they are not necessary? If one wants to find life somewhere in the Universe, one has to search for something similar to that which is proven to exist, on our planet at least. The second objection is that speculations such as these usually indicate nothing more than a poor understanding of chemistry.

It is true that chemical thinking is based on analogy, but it is also pretty dangerous to go too far with analogies. Sodium chloride, NaCl, is a white salt soluble in water, and a combination of elements analogous to sodium and chlorine, that is potassium and bromine, also produces a white salt soluble in water, KBr. Combining calcium with chlorine yields a soluble salt, $CaCl_2$, as well, but its combination with a chlorine analogue fluorine produces calcium fluoride, CaF_2, a salt very different from $CaCl_2$. Calcium fluoride is known in mineralogy as fluorite and in dentistry as a major component of tooth enamel.

The same holds true for substitution of ammonia with water or silicon with carbon. Comparison of the properties of ammonia with the properties of water, and properties of silicon with the properties of carbon is very useful in teaching chemistry, but chemistry is much more complex than any elementary course may reveal. Ammonia is definitively not water, and silicon is not carbon. Carbon dioxide, CO_2, is a gas which solidifies, turns into dry ice, at –78 °C while silicon dioxide, SiO_2, is a hard solid, known as quartz or silica, which melts at 1700 °C and boils at 2500 °C. Although there are billions of hydrocarbons, only a few of their silicon analogues, silanes, exist. It is easy to imagine an ammonia ocean full of strange silicon creatures, but we have absolutely no idea of what their biochemistry, constitution and evolution would look like. It is simply not true that "It is also true that much more attention has been paid to carbon organic chemistry than to silicon or germanium organic chemistry, largely because most biochemists we know are of the carbon, rather than the silicon or germanium variety," as Carl Sagan claimed, because this statement presupposes that there are organisms and even biochemists built of silicon or germanium compounds. It is ridiculous to propose living forms on Titan from the solitary fact that its methane ocean has an appearance superficially similar to that of an ocean on Earth. What could we possibly know about the origin of life in this ocean, when we are still waiting for a convincing explanation of the emergence of life on this planet, despite our enormous accumulated knowledge of its extant and extinct forms? Science does not rest on dreams but on evidence and productive ideas.

To put it simply, we have solid evidence for life based on water and carbon. The first proof we can build our case on is that such life exists and did evolve on Earth. The second proof is that neither water nor carbon compounds are scarce in the Universe. A productive idea would be to study their occurrence in the Solar System and planetary systems of other stars. In so doing, it is possible we won't find something green or intelligent, but we will probably find some of the most primitive forms of life such as those which existed in Hadean era on Earth. But can we prove this?

Carbon in space

That chemistry of carbon compounds is special is well known to anyone who ever studied chemistry; his first textbook was probably titled Inorganic Chemistry and the second one Organic Chemistry. The distinction between the two was even more evident in the nineteenth century when organic chemistry first came to the fore. The chemistry of substances isolated from animals and vegetables ("organized beings") appeared to be very different from those isolated from inorganic materials; that is minerals. The former are oily, combustible, and soft while the latter are usually resistant to fire and hard; upon heating, those of the first group produce coal, while those of the second transform into some kind of earth or other. In 1780 Bergman made the first distinction between organic and inorganic substances while Lavoisier noted that substances isolated from live beings produce carbon dioxide upon combustion. And finally, in his famous textbook of 1803 Berzelius divided chemistry into inorganic and organic halves, and that division persists today.

From all this it is natural to assume that bare, desolate places are not fit homes for carbon and its compounds. But that would be wrong. Geochemistry as well as astrochemistry treats carbon exactly the same as any other chemical element. The problem is not in chemistry itself, but in the level of complexity involved. In complex environments complex structures are formed and in simple environments only simple forms are present. Earth's geochemistry is much more complex than geochemistry of the Moon or asteroids; since many more rocks and minerals have been formed on Earth than on other planets, its carbon compounds are also of a much higher diversity – and Earth's carbon rocks with the most complex minerals and dynamics are usually known as living organisms.

But let us begin with the simplest examples. The Sun is primarily composed of only two elements, the lightest ones in the Universe, hydrogen and helium. But for each one hundred thousand helium atoms, or half a million hydrogen atoms, there comes one atom of carbon. The Sun contains the same number of carbon atoms as there are atoms of silicon, and carbon is ten times more abundant than sodium. In short, carbon is not a rare element in our nearest star at all. Furthermore, the nuclei of carbon atoms are catalysts of sorts – nuclear catalysts, that is. Two hydrogen nuclei do not fuse directly into a helium nucleus; instead they gradually transform into it through a series of nuclear reactions which require nuclei of carbon, nitrogen and oxygen, respectively. The first reaction in the series, known as carbon-nitrogen cycle, is the fusion of hydrogen and carbon into nitrogen. There is no sunshine without carbon. But what about some real chemistry happening in the Sun?

Due to extremely high temperatures in the Sun's atmosphere only simple carbon compounds are to be found there, such as CO and CN and, of course, carbon ions and radicals.

Chemistry of cold interstellar gas is much more complex however. In the late 1930s radicals CH, CH$^+$ and CN were detected in interplanetary space by Mt Wilson telescope, and in 1969 the first molecules, those of water and ammonia, were discovered as well. That only the simplest of molecules may be found in outer space was a common belief among astronomers of forty years ago. This was expressed by "Fisher scientific principle:" "If you cannot obtain a molecule from that well-known chemical supply house, you will not find it in the interstellar gas." But times have changed. By 2015, 184 molecules were detected in interstellar space, and many of them certainly cannot be found in Fisher's catalogue. And each year an average of six new molecules are detected.

So, where do we begin? Methane, formaldehyde, ethene, formic acid, methanol, ethanol, dimethyl-eter, acetone – and a plethora of ethyne polymers (which are certainly not to be found in Fisher's catalogue). But the most interesting substances, at least from the perspective of this book, are polycyclic aromatic hydrocarbons (PAHs), from which life on Earth presumably emerged.

The largest interstellar grains detected are 400 nm in size and contain tens of billions of atoms each. It is even difficult to decide whether we should call them particles or molecules since 400 nm is the size of both tiny particles as well as big, polymeric molecules (kerogen). At any rate, PAH molecules are very difficult to detect for two reasons. The first one is that for each molecule to be detected spectroscopically it has to have or it

Types of interstellar and circumstellar molecules

Kind	Properties	Examples
Simple ("Fisher's catalogue") organic and inorganic compounds	Stable at terrestrial conditions	HCl, CO, CH_4, c–C_6H_6, $CH_2{=}CH_2$, $HCOOCH_3$, $(CH_3)_2O$, CH_3CN, CH_3COOH, CH_3SH, $(CH_3)_2CO$, $HOCH_2CHO$, $HC{\equiv}CCHO$
Carbon chains, conjugated triple bonds	Tend to explosively polymerize	$H(C{\equiv}C)_3CN$, $CH_3(C{\equiv}C)_2CN$, $H(C{\equiv}C)_5CN$
Radicals	Very reactive species, cannot be isolated on Earth	CCH, $CCCH$, CH_3O
Carbenes and molecular ions	Very reactive species, cannot be isolated on Earth	CH_2, C_3H_2, CH^+, HCO^+, HNN^+, HCO^{2+}

has to be capable of inducing a dipole moment. Polar molecules, like those of water and ammonia (which were detected first!) all have a high dipole moment in sharp contrast to hydrocarbons, including PAHs. The second obstacle is that large molecules exhibit many spectral lines, and to measure them all is a very time-consuming job – a single spectral line may require up to 30 hours of observation.

But, against all odds, organic chemistry of interstellar "vacuum" is not poor in either variety or sheer quantity. Only 0.1% of the Universe's carbon mass is found on planets, comets and other condensed objects. All other organics reside in enormous molecular clouds, and as much as 20% of them are PAHs. "The interstellar gas contains all the raw materials of the classic Miller-Urey experiments, and some of the end products as well," wrote P. Thaddeus from Harvard University. Life really is a cosmic phenomenon.

The Martian enigma

If it were possible to ask someone in the nineteenth or early twentieth century, either an astronomer or a layman, if there was life on Mars, it is quite probable that they would give a positive answer. It is not only because invaders from H.G. Wells' novel *The War of the Worlds* hailed from that planet, but also because of the efforts of a certain American astronomer named Percival Lowell. In 1894 he built an observatory in Arizona desert to study Martian "channels" and even "oases" at which they met (both later rejected as optical illusions), convinced that they were built by Martians in order to convey fresh water from the planet's polar caps. "Let's show the Martians who's who in the Solar System," proclaimed a cartoon from those times. And let's do just that!

The first serious attempt to find life on Mars turned out to be a failure. The *Viking* mission in July 1976 performed several experiments designed to find traces of bacterial or any other life form in Martian soil, up to a depth of 10 cm, but it didn't manage to find anything. A mixture of organic compounds* did release carbon dioxide when Martian soil and water was added, which seemed promising at first (something is eating it!), though later on it was successfully explained away as an abiotic process, oxidation of formic acid by peroxides in the soil. Moreover, heating Martian soil for half a minute (sometimes repeatedly) at 200, 350 and 500 °C did not lead to detection of organic molecules. This was a great surprise, in a negative sense, of course.

The instrument tasked with detecting organics in Martian soil, a GC-MS (gas chromatograph coupled with mass spectrometer), is standard laboratory equipment today. It first separates components of a gas mixture by GS and thereafter analyzes their chemical constitution by MS. This very sophisticated and sensitive instrument did not detect any traces of organics despite the fact that meteorites rich in kerogen and other organics are regularly falling on Mars with about the same frequency as on Earth. It was estimated that some 2.4×10^5 kg of reduced carbon arrives from space to Mars each year. For the 4.5 billion year history of the Red Planet this would translate to 10^{15} kg or 10^{12} t – or roughly eight kilograms per every square meter of its surface area. Martian surface should be littered with tar, meteoritic kerogen – but nothing was found.

*Formic, lactic and glycolic acid along with amino acids glycine and alanine.

If you take a walk up a forest path on Mt Medvednica, a favorite hiking area near my native town of Zagreb, you will eventually reach a large boulder lying by the roadside and bearing a humorous inscription made by some whimsical mountaineer. It simply says "meteorite." While most passers-by would only chuckle and get on their merry way, a geologist friend of mine decided to do something about it. He took a rock sample, analyzed it and published the results in a local popular-science magazine concluding that "its meteoritic origin has not been confirmed, but it was doubtlessly proved to be a shale rock, similar to those usually found on the mountain." Scientists like to prove the obvious, but sometimes from the seemingly obvious anything but obvious may emerge. It is obvious that peroxides in the soil, generated by UV radiation, decompose organic compounds, but in spite of this a group of astronomers from the University of Florida published a paper in 2000 dealing with this particular problem. They analyzed the ways PAHs may have decomposed on Mars and found that they shouldn't have converted entirely into carbon dioxide, as other organics do, but into benzene carboxylic acids as well. They calculated that in the last three billion years, ever since the disappearance of the last fluid water on Mars, one half of a kilogram of meteorite-derived benzene carboxylic acids per square meter of Martian surface should have been formed. This should be quite a detectable amount even if this organic material was mixed up with one kilometer of Martian soil – but unfortunately this concentration wouldn't have been detectable with the *Viking's* instruments.

Whether this explanation is correct or not, there is little evidence of anything resembling a biosphere on Mars. Mars is a dry, dusty planet with a thick atmosphere (600 Pa), 95% of it being carbon dioxide. Comparison with Earth reveals that there is ten times more carbon on Mars – but only in its atmosphere. In its lithosphere, in the igneous rocks, there is about 150 times less carbon than in terrestrial rocks of the same origin (2 vs. 345 $\times$ 10^{15} t). On the other hand, it is now very clear that, once upon a time, Mars did possess a hydrosphere, liquid water running on its surface. A hydrosphere is a prerequisite for sedimentary rocks, in which 23 – 150 $\times$ 10^{15} kg of carbon, mostly in the form of carbonates, had been deposited.

It is very probable that 4.5 billion years ago the surface of Mars very closely resembled that of Earth, at the same stage of development, of course, but then it gradually lost its atmosphere. As with geological history of Earth, Martian geological history may be roughly divided into three eras, though their names are different from those of our planet's: Noachian (corresponding to terrestrial Hadean), Hesperian (similar to Archaean) and Amazonian. In the first era, Noachian, Mars was a hot planet with an atmosphere probably similar to that of Earth at the time. Gradually, plate tectonics developed and in Hesperian period a hydrosphere appeared. The most dynamic period of Mars' history was Late Noachian/Early Hesperian, with thick atmosphere and water running on its surface. If life ever existed on Mars, it probably thrived in those times, about 3.5 billion years ago.

But then it quickly disappeared again due to very rapid degassing of the planet's atmosphere. Life on Mars probably lasted less than a billion years – if it ever evolved at all.

There are no Martians on Mars. The most we can expect to find are fossils of the most primitive of organisms. But these fossils could turn out to be very valuable: they would shed a new light on the origin of life here – and elsewhere.

Follow the water

The moral of the Mars story is quite clear: you cannot grow life from carbon only. You also need liquid water. And so it is easy to write down the basic formula of life: organics + water → life. As simple as that. If life is an adaptation of organic matter to watery conditions, as was pointed out at the end of the seventh chapter, then it's immaterial from where and in which form this matter came from. Life will always find a way.

Unexpectedly, these conditions are met at a very strange place in our solar system. If you expect to find life, the places you should study are rocky planets in the so-called goldilocks zone such as Mars or Venus, and not elsewhere in the cold outer reaches of the Solar System, places like Jupiter's satellite Europa. This satellite was discovered in 1610 by Galileo, and for nearly four hundred years nobody could even imagine that this small Moon-sized celestial body might house something living… until the 1970s when ground-based astronomy detected characteristic infrared signatures of water ice on its surface. In 2000 another *Galileo*, a spacecraft this time, discovered that variations in Europa's magnetic field indicate an electrically conducting medium beneath the icy crust, presumably salty water (dissolved magnesium sulfate).* In this ocean (with a depth of 100 km and a volume of 3×10^9 km^3 – twice as much as Earth's oceans!), whose surface is hidden by kilometers of ice, hydrothermal systems rich in all kinds of organics may possibly exist, as suggested by reddish-brown filaments observed on surface ice crust. What is hiding below the surface? Perhaps some kind of radiation-driven ecosystem as proposed in 2000 by Christopher F. Chyba. This strange *life* is supposedly fed by hydrogen released by serpentinisation of the rocks on the bottom of the ocean and formaldelgde formed by radiation-driven reaction of CO_2 and H_2O on its icy surface.

These wild speculations can only be proved or disproved by a drilling mission, robotic or manned, something which is far beyond our current technology. The thick icy cover would have to be melted which would require energy from a nuclear reactor. No such device has been designed yet, at least not for use in space probes. Those designed for lengthy missions usually do possess a "nuclear source of energy," but this source is based on heat released by radioactive decay of plutonium and not by nuclear fission of uranium. This method would not be able to provide nearly enough energy to drill through a kilometer thick ice cover, not by far.

But never say never. We will visit there again in twenty or thirty years, that is virtually certain. And what will we find in that ocean? "The starfish, the horseshoe crab, the whale's backbone" or "The more delicate algae and sea anemone?" to cite T. S. Eliot again. Nothing of the kind perhaps, but finding some organic matter in a primitive stage of development is quite probable.

Follow organics, follow water and you will find life. But what kind of life? A nineteenth century comedy claims that the best answer for anything that could happen, good or bad, is *"C'est la vie,"* but I'd shy away from such a trite conclusion. A more appropriate ending to this book might be a quote from the foreword of Oparin's book *The Origin of Life*: "In 1923 I published a booklet devoted to this problem in which my views according to which life appeared in the gradual evolution of primary organic substances were first expounded." But when and where did life "appear?" Wasn't it right here, with us since the beginning of the Universe, though in less developed forms? *C'est la vie* – life is life after all.

---

*Similar structures were later discovered on other moons. Jupiter satellite Callisto has an ocean with a depth of 10 km, and probe *Cassini* discovered a geyser of water and organic molecules on the surface of Saturn's moon Enceladus. Plumes were also recently detected on Europa by Hubble Space Telescope.

Ingredients: water, organics, silicate minerals.

1 Take ingredients from the dyeing star (protoplanetary nebula).

2 Form planetesimals.

3 Arrange planetesimals into larger bodies, i.e. planets.

4 Cook for a few hundred million years until life forms.

Enjoy life!

GLOSSARY

α refers to the position next to carboxyl group, especially in the chemistry of amino acids.

abiogenesis, origin of life.

abiotic, without the aid of something living.

absolute configuration, configuration denoted (*R*) or (*S*) according to CIP rules.

accretion, growth by sticking of particles.

accretion theory states that planets grew through condensation of small dust particles into large bodies. The theory was first independently proposed by Kant and Laplace in the 18th century, and is now confirmed by astronomical observations and computer simulations.

acetic acid, the same as ethanoic acid, CH_3COOH.

acetone, the same as propanone, $(CH_3)_2CO$.

acid, a substance with dissociable hydrogen, producing hydrogen ions (H^+).

adsorption, concentration of a substance on a surface.

aerobic metabolism, metabolism which uses free oxygen (air).

Agricola, Georgius (1494 – 1555), German mineralogist, "father of mineralogy;" however, he studied Greek, Latin, Philosophy, Medicine, and Natural Sciences at universities in Germany (Leipsic) and Italy (Bologna, Venice, and probably Padua). Besides *De Re Metallica* (1556), he wrote seven works in medicine, geology, and mineralogy. He was made Burgomaster of Chemintz, Saxony, in 1546.

alanine, the simplest chiral amino acid, $CH_3CH(NH_2)COOH$, abbr. Ala, A.

aldehydes, organic compounds with general formula R–CHO.

aliphatic, containing open chain of carbon atoms.

alkanes (paraffins), saturated aliphatic hydrocarbons (C_nH_{2n+2}),
e.g. ethane, C_2H_6.

alkanoic acids, organic acids derived from alkanes, $C_nH_{2n+1}COOH$.

Altman, Sidney (b. 1939), Canadian-American biochemist. In 1989 he shared
Nobel Prize in chemistry with Thomas R. Cech.

amino acids, organic compounds with both basic amino ($-NH_2$) and acidic
carboxyl group ($-COOH$). Amino acids which build proteins (α-amino acids)
have a general formula $R-CH(NH_2)-COOH$.

amphiphilic compounds, compounds with polar (hydrophilic) and non-polar
(lipophilic or hydrohobic) parts of a molecule.

anaerobic metabolism, metabolism without free oxygen (air).

anhydride, a chemical compound which is obtained from another one by
the release of water; consequently polyphosphoric acids are anhydrides of
phosphoric acid.

animalcula, Lat. diminutive of *animal,* a tiny animal.

anoxygenic, not oxygen-producing.

apoenzyme, non-catalytic part of an enzyme.

archaea, single-celled microorganisms (prokaryotes) from kingdom Archaea;
differ from bacteria in cell membranes and metabolism (methanogenesis).

aromatic compounds, compounds with one or more aromatic rings.

aromatic ring, ring of benzene, pyridine and other unsaturated cyclic
compounds.

Arrhenius, Svante August (1859 – 1927), Swedish chemist. He founded
the theory of electrolytic dissociation and derived the first equation for
determining the rate of chemical reactions. In his book *Worlds in the Making*
(1908) Arrhenius suggested that first life came from outer space, in the form

of spores driven by stellar radiation. He was awarded the 1903 Nobel Prize in chemistry.

Asimov, Isaac (1920 – 1992), Russian-American biochemist and prolific science writer, associate professor of biochemistry at Boston University School of Medicine. He is the author of numerous popular science books as well as science fiction novels and stories.

aspartic acid, $HOCOCH_2CH(NH_2)COOH$, abbr. Asp, D.

astrobiology, the study of organic compounds in the Universe in order to find places and processes suitable for the origin of life.

asymmetry, lacking any element of symmetry. All asymmetrical objects are chiral, but not all chiral objects are asymmetrical.

autocatalysis, process in which the product of a reaction is also the catalyst for the same reaction; water is catalyst for reaction
$ZnO + 2HF \rightarrow ZnF_2 + H_2O$.

autotrophic organisms, organisms which use simple compounds, like CO_2, as a source of carbon.

B

bacillus, rod-shaped bacteria, also the name of a genus of spore-producing bacteria.

Bacillus subtilis, very common sporogeneous bacterium, obligate aerobe.

base pairing refers to the structure of DNA, where hydrogen bondings A-T and C-G stick its helices together; also present in RNA molecules (A–U, C–G).

base stacking, lipophilic interactions of nucleobases along DNA molecule.

bases, substances with a group or groups capable of binding hydrogen ions; in organic chemistry bases usually contain nitrogen atom(s), e.g. amino group $(-NH_2 + H^+ \rightarrow -NH_3^+)$.

Bellum omnia in omnes, Lat. war of everyone against everyone.

benzaldehyde, C_6H_5CHO, aldehyde of benzoic acid, C_6H_5COOH.

benzoyl, C_6H_5-CO-, abbr. Bz.

Bergman, Torbern Olof (1735 – 1784), Swedish mineralogist. He composed the first table of affinities and introduced quantitative analysis of minerals.

Berzelius, Jöns Jakob (1779 – 1848), Swedish chemist, the most influential chemist of his time. He introduced modern chemical notation (e.g. H_2O for water), composed the first reliable table of atomic weights, and proposed many terms which are still in use (catalysis, isomer, polymer, halogen, protein).

binomial nomenclature, systematic nomenclature of species, composed of two names. The first is the name of genus, and the second name is peculiar to the species; *Canis familiaris* (dog), *Canis lupus* (wolf).

biocatalyst, the same as enzyme.

biogenesis, the same as bioneogenesis, but less precise.

bioneogenesis, making of life *de novo*, origin of life.

biosynthesis, synthesis in living beings, by metabolic reaction(s).

blue-green algae and cyanobacteria are two names for the same group of organisms. They are "algae" for they are photoautrophic organisms which live in an aqueous environment; they are also "bacteria" because they are prokaryotes. Cyanobacteria are the most primitive oxygenic autotrophs, ubiquitous from Arctic ice to hot springs; symbionts with fungi in lichens.

Boscovich, Roger (1711 – 1787), Rogerius Josephus Boscovich (Lat.), or Josip Ruđer Bošković in modern Croatian spelling was a polyvalent scientist (astronomer, mathematician, physicist) as well as a geodesist, a poet and a diplomat. He was born in Dubrovnik, Croatia, but spent most of his life in Italy and France, where he held the post of the director of optical instruments in the French Royal Navy. Boscovich was a Jesuit priest until the order got temporarily suppressed in 1773.

Bubanović, Fran (1883 – 1956), Croatian biochemist and the first professor of chemistry at Zagreb University School of Medicine. Author of notable textbooks and books on chemistry and biochemistry aimed at general public.

He was a student and later a close friend of Svante Arrhenius.

buffer (solution), a solution which maintains constant pH; usually a mixture of weak acid and its salt with a strong base ($NaHCO_3/H_2CO_3$) or of a weak base and its salt with a strong acid (NH_3/NH_4Cl).

Buffon, Georges Louis Leclerc, Comte de (1707 – 1788), French naturalist. In 1739 he became the keeper of the Jardin du Roi, the French state botanical garden. During a period of fifty years, starting from 1752, he wrote 44 volumes of *Natural History*. In 1745 he suggested that Earth might have been created by collision of a massive body with the Sun (theory of catastrophe in contrast to accretion theory).

Butlerov, Aleksandr Mikhailovich (1828 – 1886), Russian chemist, active in the field of organic chemistry, especially in the light of the new structural (valence) theory.

Bütschli, Johann Adam Otto (1848 – 1920), German zoologist, professor at University Heidelberg. He studied the development of invertebrates.

C

Cahn, Robert Sidney (1899 – 1981), British chemist, editor of *Journal of Chemical Society* (1949 – 1963).

Cannizzaro, Stanislao (1826 – 1910), Italian chemist, notable for inventing a method for converting aldehydes into mixtures of alcohols and organic acids (Cannizzaro reaction). He was very eager in promoting Avogadro's ideas of molecules. Cannizzaro was also active in politics; he joined Garibaldi and became vicepresident of the Italian Senate.

capsid, coating of a virus particle (virion).

carbenes, molecular fragments with a free pair of electrons, e.g. $:CH_2$.

carbides, compounds of metals with carbon, e.g. CaC_2, Al_4C_3, Fe_3C, SiC.

carbohydrates, obsolete name for sugars, simple (monosaccharides) as well as complex ones (polysaccharides). The name simply means "hydrates of carbon;" formula of glucose, $C_6H_{12}O_6$, may be written as $C_6(H_2O)_6$.

carboxyl group (–COOH), a chemical group, characteristically acidic; $-COOH \rightarrow -COO^- + H^+$.

carboxylic acid, an acidic organic compound which contains one or more carboxyl groups.

Cech, Thomas Robert (b. 1947), American biochemist of Czech origin, professor at the University of Colorado and president of Harvard Medical Institute. He was awarded the 1989 Nobel Prize in chemistry for the discovery of ribozymes, shared with Sidney Altman.

chemoautotrophic organisms, organisms which gain energy from reactions of inorganic compounds.

Chichibabin, Aleksey Yevgenyevich (1871 – 1945) or Alexei Euguenievich Tchitchibabin was a Russian organic chemist who emigrated 1930 to France; active in heterocyclic (pyridine) and pharmaceutical chemistry.

chirality, property of an object which cannot be superimposed with its mirror image (handedness, asymmetry). Chiral objects are either left or right handed, including molecules.

chondriosomes, obsolete term for mitochondria.

chondrites, class of stony meteorites named according to chondrules, i.e. spheroidal or ellipsoidal bodies usually a few tenths of a milimeter to a few milimeters in diameter, making 70% or more mass of a chondrite; the most frequent meteorites.

chert, sedimentary rock composed of microcrystalline silica, which may be of organic (e.g. radiolaria) or inorganic origin.

chromatography, a suite of metods (paper c., column c., thin-layer c., etc.) for separation of substances, usually organic compounds by passing their solutions or vapors (gas c.) through adsorbents.

CIP rules, Cahn-Ingold-Prelog priority rules for the determination of absolute configuration, (R) or (S), of carbon and other chiral centers. They are based on the standard orientation of a molecule according to substituents at its chiral center(s).

citric acid cycle, also known as tricarboxylic acid (TCA) or Krebs cycle is a metabolic pathway for the oxidation of nutritients. They enter the cycle in the form of acetyl-coenzime A, $CH_3CO-S-CoA$.

clays, a class of argillaceous rocks which are plastic when wet; sedimentary rocks obtained by weathering of igneous rocks. By chemical compositions they are mostly alumosilicates.

clone, plant or animal formed by vegetative (asexual) reproduction, e.g. in a coral colony.

cluster, an ensemble of bound atoms or molecules that is intermediate in size between a molecule (atomic clusters) and a bulk solid (molecular clusters). It consists of $5 - 10^5$ particles.

codon, a sequence of three adjacent nucleotides that code for an amino acid. The set of all codons comprises genetic code.

coenzyme, catalytic nonprotein part of an enzyme.

coenzyme A (CoA or HS–CoA), the key coenzyme in citric acid cycle. It is composed of adenylic and pantothenic acids to which β-mercaptoethanolamine, $NH_2CH_2CH_2SH$, is bonded.

collagen, fibrous protein which on boiling yields gelatin (*kolla* in Greek).

colloids are heterogeneous mixtures consisting of a continuous phase (solid, liquid, or gas) in which tiny particles or droplets of the $1 - 1000$ nm size are dispersed. Solutions of macromolecules (e.g. proteins) are also colloids, because their size fits this range.

component, a part of a homogeneous mixture.

concentration (*c*) is expressed in mole per litre of solution (not solvent!), mol/L or similar; 1 mmol/mL = 1 mol/L, 1 μmol/mL = 1 mmol/L, etc.

condensation in organic chemistry stands for a reaction in which water is released and a new bond formed, e.g. esterification.

contact catalysis, the same as heterogeneous catalysis.

Crick, Francis Harry Compton (1916 – 2004), English biochemist, originally a physicist who turned his interest to molecular biology in order to solve the structure of DNA with his younger American colleague Watson in 1953 (the 1962 Nobel Prize in medicine and physiology). He spent the final years of his life engaged in neuroscience research.

Croatia, a European country (Republic of Croatia) on the Adriatic Sea, independent since 1991. Formerly a federal unit of the now-disbanded Yugoslavia, and a part of the Austro-Hungarian Empire prior to 1918. Population 4.5 million. During the Nazi occupation (1941 – 1945) it was declared an independent state.

cyanobacteria, the same as blue-green algae.

cyclisation, formation of cyclic molecules from acyclic ones.

cysteine, $HSCH_2CH(NH_2)COOH$, the only protein amino acid with a $-SH$ group; abbr. Cys, C.

D

Darwin, Charles Robert (1809 – 1882), English naturalist, founder of the theory of evolution. During his voyage on H. M. S. *Beagle* (1831 – 1836) he studied flora and fauna throughout the world. In 1839 he published *A Naturalist's Voyage on the Beagle*, and in 1859 his capital work *On the Origin of Species by Means of Natural Selection* followed by *The Descent of Man* (1871).

dehydrogenases, enzymes which subtract hydrogen from organic compounds by oxidizing them.

dehydrogenation, loss of hydrogen, in organic chemistry the same as oxidation, e.g. $CH_3OH \rightarrow CH_2O + 2H^+ + 2e^-$.

denaturation of proteins, process of rendering proteins biologically inactive (by heating, acidification, etc.).

deoxynucleoside, condensation product of purine or pyrimidine nucleic base with deoxyribose.

deoxyribonucleotide, a compound which consists of a purine or pyrimidine base bonded to deoxyribose, which in turn is bound to a phosphate group; phosphorylated deoxyriboside, polymer unit of DNA.

diester, double ester; it is obtained when two alcohol molecules are bonded to the same acid moiety, e.g. $2ROH + H_3PO_4 \rightarrow RO-PO(OH)-OR + 2H_2O$.

DNA, deoxyribonucleic acid, polymer of deoxyribonucleotides; substance of genes.

Drake, Frank Donald (b. 1930), American astronomer, interested especially in radioastronomy, planetary astronomy and exobiology.

Du Vigneaud, Vincent (1901 – 1978), American biochemist, professor at Cornell University Medical Colleage, New York. In 1953 he worked out the first exact order of amino acids in a natural protein oxytocine and managed to synthesize it the next year (the 1955 Nobel Prize in chemistry).

E

E. coli or *Escherichia coli*, very common bacteria widely used in genetic research. Its natural habitat is the intestine.

elimination reactions, reactions in which organic compounds turn themselves into less complex ones by releasing simple molecule(s), e.g. $CH_3CH_2OH \rightarrow CH_2{=}CH_2 + H_2O$.

enantiomer, chiral stereoisomer.

enole, alcohol with a double bond, tautomer form of an aldehyde, e.g. $CH_3CHO \leftrightharpoons CH_2{=}CHOH$.

enzyme, a catalytic protein, biocatalyst. It is usually composed of a catalytic part, coenzyme, and a polypeptide matrix, apoenzyme.

escape velocity, the velocity which a projectile or space probe need to attain to escape from a particular gravitational field. It depends on mass and diameter of heavenly body, e.g. 11.2 km/h (25,000 mph) from Earth, and 42.1 km/h (94,000 mph) from the Sun at the distance of Earth's orbit.

ester, a compound obtained by reaction of acid with alcohol, e.g. $CH_3COOH + CH_3OH \rightarrow CH_3COOCH_3 + H_2O$.

esterification, conversion of an acid into its ester.

ether, chemical compounds with a general formula R–O–R; short for diethylether, $(C_2H_5)_2O$.

eukaryotic organisms (eukaryotes, eucaryotes), organisms with cell nucleus.

evaporite, a sediment resulting from the evaporation of saline water.

exhalation, emanation of volcanic gases and vapors from the surface of Earth.

exobiology, science dealing with life forms beyond our planet (extraterrestrial life).

F

fatty acids, organic aliphatic acids, R–COOH. Biological ones have straight chains and even number of carbon atoms.

Fischer, Emil Hermann (1852 – 1919), German chemist, the forefather of protein chemistry. He also contributed to the chemistry of sugars and nucleic acids, for which he was awarded the 1902 Nobel Prize for chemistry. In 1907 he prepared the first synthetic peptides with up to 18 linked amino acids.

fluorite (fluorspar), transparent mineral, calcium fluoride, CaF_2.

Forbes, Edward (1815 – 1854), English naturalist, founder of British marine biology. He organized oceanographic expeditions to the Firth of Forth (1832), Irish Sea (1834), Orkney and Shetland Islands (1839) and Aegean Sea (1841 – 1842). He put forward the azoic hypothesis for a lifeless deep ocean.

formaldehyde, the same as methanal, HCHO.

formic acid, the same as methanoic acid, HCOOH.

formose reaction, reaction of formaldehyde in basic solutions, usually $Ca(OH)_2$, producing simple sugars.

Fox, Sidney W. (1912 – 1998), American biochemist notable for his experiment involving polymerisation of amino acids at prebiotic conditions. Since 1955 he taught at institutions in Florida and had been associated with NASA since 1960.

free radical, the same as radical.

free state, an unbonded state of an element or a compound, such as free oxygen (O_2) or free (not polymerized) amino acids.

FTIR spectra, Fourier transform infrared spectra.

G

Galileo, Galileo Galilei (1564 – 1642), Italian astronomer and physicist, best known for his defense of heliocentric theory; the first scientist to study celestial bodies with the aid of a telescope. He also planted the foundations of modern physics by solving the problem of the movement of falling bodies, the law of inertia.

gas chromatography (GC), chromatographic method for separation of substances in gas phase.

gene, fragment of DNA which encodes one protein.

genome, set of all genes in an organism.

Gilbert, Walter (b. 1932), American physicist and biochemist. He started his career as a physicist but in the 1960s he switched to biophysics and biochemistry. Gilbert was awarded the 1980 Nobel Prize in chemistry for determining the sequence of nucleotides in nucleic acids.

glyceraldehyde, $HOCH_2CH(OH)CHO$, the simplest sugar (aldotriose); oxidized glycerol.

glycerol, $HOCH_2CH(OH)CH_2OH$, or glycerin is the alcohol (triol) component of fats and oils (triglycerides).

glycine, H_2NCH_2COOH, the simplest amino acid (abbr. Gly, G); the only protein amino acid without a chiral center.

glycolic acid, $HOCH_2COOH$, the simplest hydroxy acid.

goethite, $FeO(OH)$, a mineral of blackish brown or reddish to yellow brown color. It is found in iron-rich deposits and is a product of weathering.

Gold, Thomas (1920 – 2004), Austrian-British-American astrophysicist and geophysicist. He first emigrated from Austria to England (Cambridge) and then to the United States (1956) to finally become a professor of astronomy at Cornell University. His first interest was cosmology, especially evolution of galaxies.

Graham, Thomas (1805 – 1869), Scottish physical chemist, notable as the founder of the theory of colloids. He also discovered "water of crystallisation" and studied absorption of hydrogen by palladium.

Haeckel, Ernst Heinrich (1834 – 1919), German naturalist, the first German scientist who accepted Darwinism. He was also the first to use term "ecology."

Haldane, John Burdon Sanderson (1892 – 1964), English geneticist, supporter of Oparin's theory. In 1930s he became a Communist and served as the editor of the London *Daily Worker*.

heavy metals, metals with density higher than 5 g/cm^3 such as iron and copper.

Helmont, Jan Baptist van (1577 – 1644), Flemish physician and alchemist notable for the pioneering study of gases (word "gas" was his invention; he derived it from *chaos*). Van Helmont believed in spontaneous generation as well as in transmutation of elements. He performed the first experiment in plant physiology by weighing a willow tree during its growth, thus "proving" that water was converted into wood.

heterocyclic compound, cyclic compound with a non-carbon atom (N, O or S) in the carbon ring, e.g. pyridine.

heterotrophic organisms, organisms which obtain energy by degrading organic compounds.

heterogeneous catalysis, catalysis on a solid (contact) catalyst. Reactants may be in liquid or gas phase.

high-molecular compounds, compounds with molecules of high molecular mass.

homogeneous catalysis, catalysis with a dissolved catalyst, e.g. enzyme.

Hooker, Joseph Dalton (1817 – 1911), British botanist, a close friend of Charles Darwin.

Hoyle, Fred (1915 – 2001), English astronomer, professor of astronomy at Cambridge University. He described a scheme of nuclear reactions within stellar interior (nucleosynthesis) and provided an explanation for supernova explosions.

hydration, addition of water to double or triple bond, also binding of water to a particle (ion, molecule or molecular aggregate).

hydrolysis, in organic chemistry a process contrary to condensation.

hydrophilic, in chemistry refers to molecules or their parts with polar groups (containing O, N or S) able to attract water molecules.

hydrophobic, the same as lipophilic.

hydrothermal process, any igneous activity which involves heated or superheated water and gasses (CO_2, H_2S, HF, etc.) dissolved in it.

hydroxy acids (hydroxycarboxylic acids), organic compounds with hydroxyl (–OH) and carboxyl group (–COOH), e.g. $CH_3CH(OH)COOH$ (lactic acid).

IDP, interplanetary dust particle (about 10 µm in size, composed ≈50% of kerogen).

igneous rocks, rocks which originated by cooling of magma, the primary source of material composing the Earth's surface. The most common such rocks are basalt and granite.

imidazole, $C_3H_3N_2$, aromatic heterocyclic compound with five-membered ring, abbr. Im.

Ingold, Sir Christopher Kelk (1893 – 1970), British chemist who introduced many concepts in organic chemistry, still in use today (nucleophile, electrophile, resonance effects, S_N1, S_N2, $E1$ and $E2$ reactions).

in vitro, Lat. in glass, study of biochemical reactions without participating of any living organism, e.g. enzymatic reactions in test tube. Biochemical experiments with living beings are denoted *in vivo* (Lat. in life).

IOM, insoluble organic material.

ion, electrically charged atom or molecule. Cations (e.g. Na^+, NH_4^+) are positively charged while anions (e.g. Cl^-, OH^-, $HCOO^-$) are charged negatively.

isomers, compounds with the same molecular but different structural formulae.

Kelvin (K), unit of absolute (thermodynamic) temperature; 0 K = –273.15 °C, 0 °C = 273.15 K.

Kelvin, William Thomson, Baron (1824 – 1907), Scottish mathematician and physicist; a gifted scientist who published his first mathematical paper in his teens. He discovered the minimal possible temperature (absolute zero) and put forward the second law of thermodynamics. In later life he made numerous inventions, including improvements to mariner's compasses and tide predictors.

Kendrew, John Cowdery (1917 – 1997), English biochemist notable for solving X-ray structure of hemoglobin; awarded the 1962 Nobel Prize in chemistry.

kerogen, a solid organic material which yields petroleum-type hydrocarbons through heating and distillation; high-molecular component of oil shale; also the high-molecular material from meteorites.

keto acids, acids with "ketone" (carbonyl) group, e.g. pyruvic acid, $CH_3COCOOH$.

ketones, organic compounds with general formula R–CO–R'.

Krebs cycle, the same as citric acid cycle.

Krebs, Sir Hans Adolf (1900 – 1981), German-British biochemist, notable for the discovery of citric acid cycle, in cooperation with Szent-Györgyi; awarded the 1953 Nobel Prize for physiology and medicine.

Kuckuck, Martin, a physician from St. Petersburg, Rusia, known for his book *Die Lösung des Problems der Urzeugung (Archigonia, Generatio spontanea)*, Leipzig, 1907. He was interested in the problem of insect parthenogenesis.

L

lactic acid, $CH_3CH(OH)COOH$, end product of bacterial metabolism obtained by reduction of pyruvic acid.

Lamarck, Jean Baptiste Pierre Antoine de Monet, Chavalier de (1744 – 1829), French naturalist, the first scientist who assumed that living species develop over time. In contrast to Darwin, he believed that properties of an organism acquired during its lifetime passed to its descendants. He published his theory of the inheritance of acquired characteristics (Lamarckism) in his book *Philosophie zoologique* (1809).

Lavoisier, Antoine Laurent (1743 – 1794), French chemist, "the Newton of Chemistry," founder of modern theory of combustion and the law of the conservation of mass. He also created the basis for modern chemical nomenclature.

L/D system, notation of configuration (chirality) based on comparison with the enantiomers of the simplest sugar, $HOCH_2C^*H(OH)CHO$, glyceraldehyde.

Lebedev, Pyotr Nicolaievich (1866 – 1912), Russian physicist appointed at Moscow University, notable for proving the pressure of light (1901).

Leduc, Stéphane (1853 – 1939), French biologist, professor at School of Medicine in Nantes. He described his prebiotic experiments in book *The Mechanism of Life* (1911).

Leeuwenhoek, Anton van (1632 – 1723), Dutch biologist and microscopist. He used microscopes with a single lens, but he built his instruments with such perfection that he managed to discover many secrets of the microworld. In 1683 he was the first to describe bacteria.

Liebig, Justus von (1803 – 1873), German chemist, founder of analytical organic chemistry and agrochemistry. In 1824 at University of Giessen he established the first modern method of teaching chemistry through laboratory practice. Liebig was a collaborator and a close friend to Friedrich Wöhler.

ligand, bonded atom, ion or molecule.

lignin, complex aromatic compound deposited in cell walls of plant cells making them strong and rigid. It forms 25 – 30% of tree body.

Lipman, Charles Bernard (1833 – 1944), Russian-American biologist, professor of Plant Physiology and Dean of the Graduate Division University of California. He was especially interested in soil microbes.

Lippmann, Edmund Oskar von (1857 – 1940), German chemist and natural science historian. His first interest was the chemistry of sugars.

lipophilic, in chemistry refers to molecules or their parts possessing nonpolar groups (e.g. aliphatic chains) which repulse water molecules.

Loeb, Jacques (1859 – 1924), German-American physiologist. He believed that tropism characteristic to plants might be applied to simple animals. In 1899 he showed that unfertilized sea-urchin egg could be made to develop to maturity by proper environmental changes ("artificial parthenogenesis").

Lowell, Percival (1855 – 1916), American astronomer notable for his study of "Martian canals," for which he even built a specialist observatory in Arizona (Lowell Observatory, opened in 1894). Instead of discovering Martians he found a new planet – Pluto.

low-molecular compounds, compounds with molecules of low molecular mass.

LUCA (also LCA, LUA), the last universal common ancestor, hypothetical organism from which all life forms on Earth evolved; probiont, progenote.

M

macromolecular, composed of big molecules, presumably polymers or copolymers.

macromolecules, the same as biopolymers – polysaccharides, proteins and nucleic acids.

magmatic rock, igneous rock formed by magmatic differentiation, such as fractional crystallization.

magnetite, Fe_3O_4, opaque mineral with submetallic luster, used as iron ore; may be magnetic (lodestone).

mass spectroscopy (MS), method of instrumental analytical chemistry by which a substance is first ionized and decomposed, and then its fragments (ions) are separated in a magnetic field. Substances are identified by m/e signals of their ions.

Maxwell, James Clerk (1831 – 1879), Scottish mathematician and physicist known for the classical theory of electromagnetic radiation and kinetic theory of heat. In his youth Maxwell demonstrated that the rings of Saturn are not liquid or gaseous, as was commonly believed, but instead consist of myriads of small bodies.

Mendeleev, Dmitri Ivanovich (1834 – 1907), Russian chemist who is best known as the discoverer of the periodic systems of the elements (1869) although he was also active in geochemistry, launching a hypothesis of inorganic origin of petroleum and proposing the gasification of coal *in situ*.

mer, polymeric unit, e.g. CH_2 in polyethylene molecule, $(-CH_2-)_n$.

meteor, atmospheric phenomenon accompanying the fall of a meteorite; shooting star.

meteorite, extraterrestrial material which falls to the surface of Earth. Meteorites are differentiated as iron (siderites), stony (stones, aerolites), stony-iron (siderolites, pallasites), and glassy meteorites (tektites).

methanogenesis, production of methane by microbial (archaea) metabolism.

methionine, $CH_3S(CH_2)_2CH(NH_2)COOH$, sulfur-containing protein amino acid, abbr. Met, M.

micelle, an aggregate of surfactant (amphiphilic) molecules dispersed in a liquid colloid.

micrometer (μm), unit of length equal to 10^{-6} m; 1 μm = 0.001 mm.

Miller, Stanley Lloyd (1930 – 2007), American chemist notable for synthesis of amino acids in prebiotic conditions in 1953. He achieved this while still being a postgraduate student of Urey. Later he became interested in life on Mars and participated in the *Viking* project.

mitochondria, microscopic bodies occurring in cytoplasm, up to 2,500 per cell, 0.5 μm in diameter; centers of citric acid cycle, i.e. oxidative metabolism.

mole (mol), Avogadro number (N_A = 6.022 × 10^{23}) of particles (atoms, molecules, electrons, etc.).

molecular ions, ions derived from molecules, not atoms.

molecular mass, short for relative molecular mass, i.e. mass of a molecule relative to 1/12 mass of carbon isotope ^{12}C.

molecular weight (MW), obsolete for molecular mass.

mRNA, messenger ribonucleic acid or messenger RNA. It serves as a template for protein synthesis on ribosomes.

myristic acid, $CH_3(CH_2)_{12}COOH$, saturated fatty acid found in coconut oil and butter fat.

N

NAD, nicotinamide adenine dinucleotide. It exists in two forms, NADH (reduced) and NAD^+ (oxidized). It is a common coenzyme of dehydrogenases.

NADP, nicotinamide adenine dinucleotide phosphate, phosphorylated NAD.

nanometer (nm), unit of length equal to 10^{-9} m; 1 nm = 0.001 μm.

naturally occurring amino acids, vague term for protein amino acids.

NMR spectroscopy, nuclear magnetic resonance spectroscopy, standard method used for determination of structure (constitution) of molecules.

nucleoside, ribonucleoside or deoxyribonucleoside.

nucleotide, ribonucleotide or deoxyribonucleotide.

O

Oparin, Aleksandr Ivanovich (1894 – 1980), Russian biochemist, professor of plant biochemistry at Moscow University; the founder of the modern theory of the origin of life, as presented in his book *The Origin of Life on Earth* (1936).

organic acids usually have a carboxyl group as their dissociable group: $-COOH \rightarrow -COO^- + H^+$.

Orgel, Leslie Eleazer (1927 – 2007), British chemist affiliated to University of Oxford and University of Cambridge. In addition to the origin of life, Orgel also had scientific interest in transition metal chemistry.

Ostwald, Friedrich Wilhelm (1853 – 1932), Russian-German physical chemist. He was interested in catalysis (the 1909 Nobel Prize in chemistry) and philosophy of science.

oxidation, chemical process in which electrons are released, e.g. $Cu^+ \rightarrow Cu^{2+} + e^-$.

oxidation state refers to the respective element. The lowest oxidation state (–I) of chlorine is in chlorides (Cl^-), and the highest (+VII) in perchlorates (ClO^-_4). Accordingly, the redox reactions in organic chemistry could be viewed as a change of oxidation state of carbon; starting from the highest (+IV), i.e. CO_2, gradually to the lowest: CO (+II), CH_2O (0), CH_3OH (–II), CH_4 (–IV).

oxidative or oxidizing (state, atmosphere, conditions, etc.) refers to the capacity of oxidation.

oxidative (oxidizing) atmosphere, an atmosphere rich in oxidizing substances, such as free oxygen and its radicals.

oxygenic organisms, those which produce free oxygen.

ozone, O_3, a very reactive allotropic form of oxygen, O_2.

P

Page, David (1814 – 1879), English geologist, professor at Durham University College of Physical Sciences and a prolific writer.

PAH, polycyclic aromatic hydrocarbon(s); aromatic hydrocarbons with fused aromatic rings such as naphthalene, anthracene, pyrene or coronene. Many of them are cancerogenic.

Paracelsus (1493 – 1541) is a nickname which Phillipus Aureolus Theophrastus Bombastus von Hohenheim gave to himself, meaning "greater than Celsus." This Swiss-born physician and alchemist tried to reform medicine by accepting Arabic and refuting Roman (Galen) influence. In alchemy he adopted the teaching of *tria prima* (mercury, sulfur, salt), instead of the four Greek elements (water, air, earth and fire). He insisted on medicines being prepared from minerals, instead of herbs, founding iatrochemistry, from which both modern chemistry and pharmacy ultimately developed.

Pasteur, Louis (1822 – 1895), French chemist, discovered "molecular asymmetry" (chirality) in 1848. Later he studied fermentation, which led him to microbiology and eventually to immunology (anthrax and rabies vaccines).

Pauling, Linus Carl (1901 – 1994), American chemist, who was awarded two Nobel Prizes, the first for chemistry (1954) and the second for peace (1963). He introduced the new concept of chemical bonding, particularly in his textbook *The Nature of the Chemical Bond* (1939). In 1970 he started to popularize a hypothesis that large doses of vitamin C could help fight many illness, including cancer.

peptide bond, –CO–NH–, a bond by which amino acids are linked into peptide chain.

peptides, polymers of amino acids via peptide bond; functional peptides with many polymer units (polypeptides) are known as proteins.

Perutz, Max Ferdinand (1914 – 2002), Austrian-British biochemist. In 1953 he solved the structure of myoglobin at Cambridge University. He shared the 1962 Nobel Prize in chemistry with Kendrew.

pH, a well-known scale of acidity defined as negative logarithm of the concentration of hydrogen ions, $pH = -\lg[H^+]$. Solutions with pH < 7 are acidic, those with pH = 7 are neutral, and those with pH > 7 are basic.

phase in chemistry is a part of heterogeneous mixture.

phosphine, hydride of phosphorus, PH_3.

phosphoric acid, H_3PO_4 or $(HO)_3PO$, is only one of many phosphoric acids (orthophosphoric acid). It forms condensation polymers at 200 – 300 °C, and by heating at high temperatures it is converted into metaphosphoric acid, HPO_3.

phosphorylation, formation of esters of phosphoric acid.

photic zone, surface zone of sea or lake sufficiently illuminated for photosynthesis.

photosynthesis, biosynthesis by using the energy of light.

phototrophic organisms, organisms which use light as a source of energy for synthesis of complex organic compounds; autotrophic organisms dependent on light.

Piccard, Auguste (1884 – 1962), Swiss physicist, affiliated to Polytechnic Institute in Brussels from 1922 to his retirement in 1954. He studied cosmic rays but is much better known for his expeditions into the unknown; in 1931 he used a balloon to ascend to an altitude of 51,775 ft (15,781 m), and in 1930s he constructed bathyscaphe *Trieste* which ultimately reached the bottom of the ocean (the Marianas Trench) at the depth of 36,204 ft (11,035 m) in 1960.

pioneer organism, "a chemically deterministic entity at the threshold between abiotic and biotic world" (Wächtershäuser).

plastids, bodies in cytoplasm of plant cells, containing reserve food materials and/or pigments (e.g. chlorophyll in chloroplasts).

polar group, a group of atoms in a molecule exerting dipole moment, e.g. –OH, –COOH, –NH$_2$.

polar molecules, molecules with dipole moment, e.g. water, ammonia and ethanol.

Ponnamperuma, Cyril (1923 – 1994), Sri Lankan-American biochemist, at University of California since 1962. He used Miller-type experiments for the synthesis of nucleic acid components.

potassium ferrocyanide, $K_4[Fe(CN)_6]$, or potassium hexacyanoferrate(II), according to new (IUPAC) nomenclature, is a yellow powder used in medicine and color industry.

ppb, parts per billion, $\mu g/kg$ or mg/t; 1 ppb = 0.001 ppm.

ppm, parts per million, $\mu g/g$, mg/kg, or g/t; 1 ppm = 0.001‰.

prebiotic, before the emergence of life.

Prelog, Vladimir (1906 – 1998), Croatian chemist (the 1975 Nobel Prize in chemistry) since 1941 professor at Organic Chemistry Laboratory at ETH, Zurich, Switzerland; notable for developing the chemistry of compounds with chiral centers, and especially for CIP rules.

prokaryotic organisms (prokaryotes, procaryotes) organisms whose genome (DNA) is not separated by membrane from cytoplasm, that is organisms without cell nucleus. They also lack mitochondria and chloroplasts.

protein amino acids, or naturally occurring amino acids, is a set of 20 L-α-amino acids from which proteins are built.

proteins, biologically functional polymers of α-amino acids, but frequently with other moieties (sugars, porphyrins, metals, etc.). They usually have about 100 – 200 monomer units.

proteinoids, polymers of amino acids resembling proteins.

PVC, poly(vinyl-chloride), $(-CH_2-CHCl-)_n$, polymer of vinyl-chloride, $CH_2=CHCl$, a common type of plastics.

pyrite, FeS_2, opaque mineral of metallic luster and pale yellow color nicknamed "fool's gold," crystallized in a cubic system. It is found in igneous, sedimentary and metamorphic rocks.

pyruvic acid, $CH_3COCOOH$, a key metabolite in aerobic and anaerobic metabolisms.

R

R in chemical formulas denotes a hydrocarbon chain ("radical"), also the absolute configuration of a chiral center, (*R*).

racemate, mixture of equal parts of (*R*) and (*S*) enantiomer.

radical, or free radical, molecular species with unpaired electron(s), e.g. HO (HO·). Radicals are very reactive (reactions of free radicals).

radical theory, a 19th century theory of chemical constitution based on an assumption that some parts of organic molecules (radicals) are chemically unchangeable and behave as chemical elements in inorganic chemistry. Accordingly, ethanol was written as $C_2H_4 \cdot H_2O$ and chloroethane as $C_2H_4 \cdot HCl$.

Redi, Francesco (1626 – 1697), Italian physician and poet, chiefly known for *Bacco in Toscana*. In 1668 he made the first experiment opposing the theory of spontaneous generation.

redox reactions, reaction in which electrons are interchanged, e.g. $Fe + Cu^{2+} \rightarrow Fe^{2+} + Cu$. In organic chemistry and biochemistry they could also be seen as hydrogen transfer reactions, e.g. $C_6H_{12}O_6 + 6O_2 \rightarrow 6CO_2 + 6H_2O$.

reduced carbon, elementary carbon or any carbon compound containing carbon in the oxidative state < 0, such as hydrocarbons; organic compounds in general.

reducing (state, atmosphere, conditions, etc.) refers to a capacity for reduction.

reducing atmosphere, atmosphere rich in reductive compounds, such as free hydrogen and methane.

reduction, chemical process in which electrons are accepted, e.g. $Fe^{3+} + e^- \rightarrow Fe^{2+}$.

reductive citric acid cycle, reverse TCA cycle or Arnon cycle, a metabolitic pathway for CO_2 fixation in autotrophic organisms.

replication, doubling of DNA or hypothetical pre-RNA molecule.

ribonucleoside, condensation product of purine or pyrimidine nucleic base with ribose.

ribonucleotide, a compound which consists of a purine or pyrimidine base bonded to ribose, which is in turn bound to a phosphate group; phosphorylated nucleoside. Ribonucleotides are polymer units of RNA.

ribosomes, granules of proteins and RNA (rRNA) present in cytoplasm of all types of organisms, about 15 nm in diameter. By binding to mRNA and tRNA they synthesize proteins. In so doing they are arranged in polyribosomes along mRNA molecules.

Rich, Alexander (1924 – 2015), American biochemist and biophysicist, notable for the discovery of Z modification of DNA molecule.

Richter, Hieronymus Theodor (1824 – 1898), German chemist and mineralogist, affiliated at the Freiberg School of Mines; co-discoverer of element indium.

RNA, ribonucleic acid, polymerized ribonucleotides.

Ross, John (1777 – 1856), British naval officer and Arctic explorer. He led three Arctic expeditions (1818, 1829 – 1833, 1850) in order to find the Northwest Passage.

rRNA, ribosomal RNA, a constituent of ribosomes.

Ruzicka, Leopold (1887 – 1976), or Ružička in Croatian spelling, was a Croatian-Swiss chemist, notable for the discovery of isoprene rule for biosynthesis of natural products. He was awarded the 1939 Nobel Prize in chemistry.

S

Sagan, Carl (1934 – 1996), American astronomer, since 1968 associate professor at Cornell University and director of its Laboratory for Planetary Study. Author of many popular books in astronomy. He was especially interested in exobiology and trying to find traces of life on other planets.

Sanger, Frederick (1918 – 2013), English biochemist at Cambridge University; worked in the same laboratory as Kendrew, Perutz, Watson, and Crick. In 1953 he determined amino acid sequence of insulin and was awarded the 1958 Nobel Prize in chemistry and, once again, in 1980 for the invention of a new method of DNA sequencing.

schreibersite, $(Fe,Ni)_3P$, a major phosphoric mineral in meteorites.

Schrödinger, Erwin (1887 – 1961), Austrian physicist (the 1933 Nobel Prize in physics), one of the forefathers of quantum physics (Schrödinger wave equation). In 1938 he immigrated to England and returned to Vienna, Austria in 1956.

sequence, linear order of amino acids in a peptide or protein (amino acid sequence) or nucleotides in DNA or RNA (nucleotide sequence).

serine, $HOCH_2CH(NH_2)COOH$, protein amino acid with a polar side chain, abbr. Ser, S.

serpentinisation, a process of weathering of silicate minerals, usually olivine, $(Mg,Fe)_2SiO_4$, in which serpentine (a group of at least 16 minerals), $(Mg,Fe,Ni)_3Si_2O_5(OH)_4$, is formed. The hydrogen gas is frequently released due to oxidation of iron(II):
$3Fe_2SiO_4 + 2H_2O \rightarrow 3SiO_2 + 2Fe_3O_4 + 2H_2$.

silanes, compounds of hydrogen with silicon (SiH_4, Si_2H_6, etc.); gases or liquids which begin to burn in the contact with air.

stereoisomers, isomers with the same constitutional formula but different projection formulas, e.g. enantiomers.

stromatolite, a laminated sedimentary structure formed by microorganisms, especially cyanobacteria in shallow water.

sugars, simple sugars or monosaccharides, is common name for polyalcohols possessing a keto (ketoses) or aldehyde functional group (aldoses). According to the number of carbon atoms they are differentiated as trioses (C_3), tetroses (C_4), pentoses (C_5) and hexoses (C_6); glucose is aldohexose and ribose is ketopentose. By cyclyzation they form five- (furanosides) and six-membered rings (pyranosides).

Szent-Győrgyi, Albert (1893 – 1986), Hungarian-American biochemist notable for isolation of vitamin C (1932), which earned him the 1937 Nobel Prize in medicine and physiology. He also isolated vitamin P (1936) and did an important work in elucidation of citric acid cycle. In 1947 he immigrated to the United States and became an American citizen in 1955.

T

tektite, a glassy meteorite.

thioethers, chemical compounds with a general formula R–S–R, organic sulfides.

thiols, organic compounds with a general formula R–SH, also known as thioalcohols or mercaptans.

Tiselius, Arne Wilhelm Kaurin (1902 – 1971), Swedish chemist, notable for the invention of Tiselius U tube for separating colloids, including proteins, by electrophoresis (1937). He was awarded the 1948 Nobel Prize in chemistry.

transamination, reaction of conversion of amino acids into keto acids, and *vice versa*, by interchanging O and N atoms; very common in the biosynthesis of amino acids.

transcription, synthesis of mRNA sequence according to sequence of DNA.

translation, the process of polypeptide chain synthesis according to sequence of mRNA.

Traube, Moritz (1826 – 1894), German chemist who worked in Ratibor, Breslau and Berlin in the field of physiological chemistry.

tricarboxylic acid cycle, TCA, the same as citric acid cycle.

Trier, Georg (1884 – 1944), German chemist born in Prague and engaged mostly in the chemistry of natural products. In 1910 he defended his PhD thesis at ETH, Zurich, and wrote an influential book *Chemie der Pflantzstoffe* (1924).

tRNA, transfer ribonucleic acid or transfer RNA, a small RNA molecule resembling a protein in many ways. It binds amino acids in the process of protein synthesis.

Tsiolkovsky, Konstantin Eduardovich (1857 – 1935), Russian physicist, a pioneer of space flight. In 1903 he began writing a series of articles for an aviation magazine on the possibilities of space flight and later wrote a science fiction novel on the subject (*Outside the Earth*). He derived the first physical formula for rocket flight.

U, V

Urey, Harold Clayton (1893 – 1981), American chemist; awarded the 1943 Nobel Prize for the discovery of heavy hydrogen, deuterium (1931). Later on he developed detailed theories of planetary formation.

Van't Hoff, Jacobus Hendricus (1852 – 1911), Dutch physical chemist, notable for the tetrahedral model of carbon atom. He was the first chemist awarded with the Nobel Prize (1901).

vesicule (Lat. *vesicula*, bubble), a bubble formed by encapsulation of liquid by amphiphilic membrane, a protocell.

Virchow, Rudolf Ludwig Karl (1821 – 1902), German pathologist, professor of pathological anatomy at Berlin University. He was among the first to accept the cell theory, though he rejected Pasteur's germ theory of disease. He was also the first to describe leukemia (1845).

W

Watson, James Dewey (b. 1928), American biochemist notable for the discovery of "double helix" of DNA in 1953 (the 1962 Nobel Prize in physiology and medicine). Watson was a "quiz kid"; he entered University of Chicago at fifteen and graduated at nineteen. Being a fluent writer he described the discovery in his book *The Double Helix* (1968).

weathering, a process by which rocks are broken down and decomposed by action of external agencies such as wind, rain, temperature changes, plants, and bacteria.

Weismann, August Friedrich Leopold (1834 – 1914), German biologist, professor of zoology at University Freiburg-im-Breisgau. To prove Darwin's theory of evolution, he cut off the tails of 1592 mice to show that they all produced offspring with full-sized tails.

Westheimer, Frank Henry (1912 – 2007), American chemist affiliated at Harvard University. He was active in the field of stereochemistry of biochemical processes.

Wöhler, Friedrich (1800 – 1882), German chemist, student of Berzelius and a close friend to Liebig. He is notable for the first organic synthesis (1828), by conversion of ammonium cyanate into carbamide (urea),
$NH_4OCN \rightarrow (NH_2)_2CO$.

Wurtz, Charles (1817 – 1884), French chemist notable for the discovery of reaction for elongation of aliphatic chains (Wurtz reaction).

X, Y, Z

X-ray crystallography (X-ray diffraction analysis), the use of diffraction patern produced by X-ray scattering from crystals to determine the 3D structure of molecules.

Yugoslavia, federal European state in western Balkans founded after the First World War; at first a monarchy ruled by a Serbian dinasty (1918 – 1941) and then a Communist-ruled republic (1945 – 1991). It was consisted of six major nations which formed independent states following Yugoslavia's dissolution (Slovenia, Croatia, Bosnia-Herzegovina, Serbia, Montenegro, Kosovo, and Macedonia).

Zagreb, the capital of Croatia, approximately 700,000 inhabitants; situated near Mt Medvednica and the Sava River.

zeolites, a group of tectosilicates containing true water of crystallisation. They are found in volcanic rocks and hydrothermal vents.

zircon, a zirconium mineral ($ZrSiO_4$); superb gem approaching diamond in fire and brilliance.

LITERATURE

There are many books and papers, from the very scientific to those intended for popular audience, concerning the origin of life. Only the most important ones and those referring directly to the data in the respective chapter are listed here.

Chapter 1

Ch. Darwin, *The Origin of Species*, Edited with an Introduction and Notes by Gilliam Beer, Oxford World's Classics, Oxford Univ. Press, Oxford, 1996 (Ch. Darwin, *On The Origin of Species By Means of Natural Selection, or the Preservation of Favoured Races in the Struggle for Life*, 2nd Ed., London, 1859).

D. Deamer, S. Singaram, S. Rajamani, V. Kompanichenko, S. Guggenheim, Self-assembly processes in the prebiotic environment, *Phil. Trans. Roy. Soc. B* **361** (2006) 1809–1818, https://doi.org/10.1098/rstb.2006.1905.

J. Farley, Philosophical and historical aspects of the origin of life, *Treb. Soc. Cat. Biol.* **39** (1986) 37–47.

J. P. Ferris, A. R. Hill, R. Liu, L. E. Orgel, Synthesis of long prebiotic oligomers on mineral surfaces, *Nature* **381** (1996) 59–61, https://doi.org/10.1038/381059a0.

A. I. Oparin, Theories of spontaneous generation of life, in: *The Origin of Life*, translated by S. Morgulis, 2nd Ed., Dover Publ., Mineola, New York, 1953, pp. 1–28.

J. Peretó, J. L. Bada, A. Lazcano, Charles Darwin and the origin of life, *Orig. Life Evol. Biosph.* **39** (2009) 395-406, https://doi.org/10.1007/s11084-009-9172-7.

N. Raos, V. Bermanec, Catalysis in the primordial world, *Kem. Ind.* **66**(11-12) (2017) 641–654, https://doi.org/10.15255/KUI.2017.014.

Chapter 2

P. Clancy, A. Brack, G. Horneck, *Looking for Life, Searching the Solar System*, Cambridge Univ. Press, New York, 2005.

R. A. Kerr, Ancient life on Mars?, *Science* **273** (1996) 864–866, https://doi.org/10.1126/science.296.5572.1384.

J. Love, The Martian fossils, http://www.synapses.co.uk./science/martian.html, p. 1-6, (release January 15, 1998).

C. Meyer, Mars Meteorite Compendium, http://www-curator.jsc.nasa.gov/curator/antmet/mmc/mmc.htm.

E. G. Nisbert, N. H. Sleep, The habitat and nature of early life, *Nature* **409** (2001) 1083–1091, https://doi.org/10.1038/35059210.

D. F. Salisbury, Meteorite yields evidence of primitive life on early Mars, Stanford News Service, http://news.stanford.edu/pr/96/960806marsfossil.htlm, pp. 1–3, (release from July 8, 1996).

N. H. Sleep, K. J. Zahnle, J. F. Kasting, H. J. Morowitz, Annihilation of ecosystems by lagre asteroid impacts on the early Earth, *Nature* **342** (1989) 139–142, https://doi.org/10.1038/342139a0.

T. Stephan, E. K. Jessberger, C. H. Heiss, D. Rost, TOF-SIMS analysis of polycyclic aromatic hydrocarbons in Allan Hills 84001, *Met. Planet. Sci.* **38**(1) (2003) 109–113, https://doi.org/10.1111/j.1945-5100.2003.tb01049.x.

P. Weber, J. M. Greenberg, Can spores survive in interstellar space?, *Nature* **316** (1985) 403–407, https://doi.org/10.1038/316403a0.

J. Westman, Is there life on Mars?, http://www.ufo.se/ufofiles/english/issue 3/marsfoss.htm, pp. 1–5 (release from Sept. 9, 2010).

Chapter 3

L. M. Barge et al., From chemical gardens to chemobrionics, *Chem. Rev.* **115** (2015) 8652-8703, https://doi.org/10.1021/acs.chemrev.5b00014.

U. Deichmann, "Molecular" versus "colloidal": controversies in biology and biochemistry, 1900 – 1940*, *Bull. Hist. Chem.* **32**(2) (2007) 105–118.

C. Del Bianco, S. F. Mansy, Nonreplicating protocells, *Acc. Chem. Res.* **45** (2012), https://doi.org/10.1021/ar300097w.

J. B. S. Haldane, The origin of life, *Rationalist Annu.* **148** (1929) 3–10.

M. S. Kritsky, Momoirs of Aleksandr Ivanovich Oparin, *Appl. Biochem. Microbiol.* **41**(3) (2005) 316–318; translated from *Prikl. Biokhim. Mikrobiol.* **41** (2005) 358–360.

S. L. Miller, J. W. Schopf, A. Lazcano, Oparin's "Origin of Life": sixty years later, *J. Mol. Evol.* **44** (1997) 351–353, https://doi.org/ 10.1007/PL00006153.

A. I. Oparin, *The Origin of Life*, transl. by S. Morgulis, 2nd Ed., Dover Publ., Mineola, New York, 1953.

N. Raos, Carbide chemistry and Oparin's theory on the origin of life, *Bull. Hist. Chem.* **42**(1) (2017) 57–62.

R. Robinson, The origins of petroleum, *Nature* **212** (1966) 1291–1295, https://doi.org/10.1038/2121291a0.

Chapter 4

J. L. Bada, A. Lazcano, Stanley L. Miller 1930 – 2007, *Biographical Memoirs*, National Academy of Sciences, http://www.nasonline.org/memoirs.

J. L. Bada, A. Lazcano, Prebiotic soup – revisiting the Miller experiment, *Science* **300** (2003) 745-746, https://doi.org/10.1126/science.1085145.

H. J. Cleaves II, The prebiotic chemistry of formaldehyde, *Precamb. Res.* **164** (2008) 111-118, https://doi.org/10.1016/j.precamres.2008.04.002.

J. P. Ferris, C. T. Chen, Chemical evolution. XXVI. Photochemistry of methane, nitrogen, and water mixtures as a model for the atmosphere of the primitive Earth, *J. Amer. Chem. Soc.* **97** (1975) 2962–2967, https://doi.org/10.1021/ja00844a007.

S. W. Fox, The chemical problem of spontaneous generation, *J. Chem. Educ.* **34** (1957) 472–479, https://doi.org/10.1021/ed034p472.

S. W. Fox, K. Harada, Thermal copolymerization of amino acids to a product resembling protein, *Science* **128** (1958) 1214, https://doi.org/10.1126/science.128.3333.1214.

K. Harada, S. W. Fox, The thermal condensation of glutamic acid and glycine to linear peptides, *J. Amer. Chem. Soc.* **80** (1957) 2694–2697, https://doi.org/10.1021/ja01544a027.

S. L. Miller, A production of amino acids under possible primitive Earth conditions, *Science* **117** (1953) 528-529, https://doi.org/10.1126/science. 117.3046.528.

S. L. Miller, Production of some organic compounds under possible primitive Earth conditions, *J. Amer. Chem. Soc.* **77** (1955) 2351–2361, https://doi.org/10.1021/ja01614a001.

N. Neal-Jones, B. Stiegerwald, "Lost" Miller experiment gives pungent clue to origin of life, Goddard Space Flight Center, http://www.nasa.gov/centers/goddard/news/releases/2011/lost_exp.htmlrelease.

D. L. Rohlfing, Thermal polyamino acids: synthesis at less than 100 ºC, *Science* **193** (1976) 68-70, https://doi.org/10.1126/science.935858.

N. Wade, Stanley Miller, who examined origins of life, dies at 77, *New York Times*, May 23, 2007 http://www.nytimes.com/2007/05/23/us/miller.html.

Chapter 5

I. V. Delidovich, A. N. Simonov, O. P. Taran, V. N. Parmon, Catalytic formation of monosaccharides: from the formose reaction towards selective synthesis, *Chem. Sus. Chem.* **7** (2014) 1833–1846, https://doi.org/10.1002/cssc.201400040.

J. P. Ferris, Montmorillonite-catalysed formation of RNA oligomers: the possible role of catalysis in the origins of life, *Phil. Trans. Roy. Soc. B* **361** (2006) 1777–1786, https://doi.org/10.1098/rstb.2006.1903.

W. Gilbert, The RNA world, *Nature* **319** (1986) 618, https://doi.org/10.1038/319618a0.

R. M. de Graaf, J. Visscher, A. W. Schwartz, A plausibly prebiotic synthesis of phosphonic acids, *Nature* **378** (1995) 474-477, https://doi.org/10.1038/378474a0.

G. J. Handschuh, L. E. Orgel, Struvite and prebiotic phosphorylation, *Science* **179** (1973) 483-484, https://doi.org/10.1126/science.179.4072.483.

G. F. Joyce, RNA evolution and the origins of life, *Nature* **338** (1989) 217–224, https://doi.org/10.1038/338217a0.

R. Lohrmann, L. E. Orgel, Urea-inorganic phosphate mixtures as prebiotic phosphorylating agents, *Science* **171** (1971) 490–494, https://doi.org/10.1126/science.171.3970.490.

R. E. Nielsen, Peptide nucleic acid (PNA): a model structure for the primordial genetic material?, *Orig. Life Evol. Biosph.* **23** (1993) 323–327, https://doi.org/10.1007/BF01582083.

P. Nissen, J. Hansen, N. Ban, P. B. Moore, T. A. Steitz, The structural basis of ribosome activity in peptide bond synthesis, *Science* **289** (2000) 920–930, https://doi.org/10.1126/science.289.5481.920.

J. Rabinowitz, S. Chang, C. Ponnamperuma, Phosphorylation by way of inorganic phosphate as a potential prebiotic process, *Nature* **218** (1968) 442–443, https://doi.org/10.1038/218442a0.

A. Ricardo, Bioorganic molecules in the cosmos and the origin of Darwinian molecular systems, PhD Thesis, University of Florida, 2004.

A. Ricardo, M. A. Carrigan, A. N. Olcott, S. A. Benner, Borate minerals stabilize ribose, *Science* **303** (2004) 196, https://doi.org/10.1126/science.109464.

A. Ricardo, J. W. Szostak, Origin of life on Earth. Fresh clues hint at how the first living organisms arose from inanimate matter, *Sci. Amer.* **301**(3) (2009) 54–61, https://doi.org/10.1038/scientificamerican0909-54.

C. Sanson, When is an enzyme not a protein? When it's a ribozyme, *Chem. World* **13** (2016) 62–65.

A. W. Schwartz, Phosphorus in prebiotic chemistry, *Phil. Trans. Roy. Soc. B* **361** (2006) 1743–1749, https://doi.org/10.1098/rstb.2006.1901.

S. G. Srivatsan, Modeling prebiotic catalysis with nucleic acid-like polymers and its implications for the proposed RNA world, *Pure Appl. Chem.* **76** (2004) 2085–2099, https://doi.org/10.1351/pac200476122085.

W. R. Taylor, Stirring the primordial soup. RNA world: does changing the direction of replication make RNA life viable?, *Nature* **434** (2005) 705, https://doi.org/10.1038/434705a.

W. R. Taylor, Transcription and translation in an RNA world, *Phil. Trans. Roy. Soc. B* **361** (2006) 1751-1760, https://doi.org/10.1098/rstb.2006.1910.

J. D. Watson, F. H. C. Crick, A structure of deoxyribose nucleic acid, *Nature* **421** (1953) 737–378, https://doi.org/10.1038/171737a0.

J. D. Watson, *Molecular Biology of the Gene*, W. A. Benjanim, Inc., New York, 1970.

F. H. Westheimer, Polyribonucleic acids as enzymes, *Nature* **319** (1986) 534–536, https://doi.org/10.1038/319534a0.

F. H. Westheimer, Why nature chose phosphates, *Science* **235** (1987) 1173–1178, https://doi.org/10.1126/science.2434996.

Y. Yamagata, H. Watanabe, M. Saitoh, T. Namba, Volcanic production of polyphosphates and its relevance to prebiotic evolution, *Nature* **352** (1991) 516–519, https://doi.org/10.1038/352516a0.

A. J. Zaug, T. R. Cech, The intervening sequence RNA of *Tetrahymena* is an enzyme, *Science* **231** (1986) 470–475, https://doi.org/10.1126/science.3941911.

Chapter 6

T. R. Anderson, T. Rice, Deserts on the sea floor: Edward Forbes and his azoic hypothesis for a lifeless deep ocean, *Endeavour* **30** (2006) 131–137, https://doi.org/10.1016/j.endevour.2016.10.003.

J. A. Baross, S. E. Hoffman, Submarine hydrothermal vents and associated gradient environments as sites for the origin and evolution of life, *Orig. Life* **15** (1985) 327–345, https://doi.org/10.1007/BF01808177.

T. Gold, The deep, hot biosphere, *Proc. Natl. Acad. Sci. USA* **89** (1992) 6045-6049.

D. O. Hall, R. Cammack, K. K. Rao, Role for ferredoxins in the origin of life and biological evolution, *Nature* **233** (1971) 136–138, https://doi.org/10.1038/233136a0.

R. Lill, Function and biogenesis of iron-sulphur proteins, *Nature* **460** (2009) 831–838, https://doi.org/10.1038/nature08301.

R. Österberg, Origins of metal ions in biology, *Nature* **249** (1974) 382–383, https://doi.org/10.1038/249382a0.

D. C. Rees, J. B. Howard, The interface between the biological and inorganic worlds: iron-sulfur metalloclusters, *Science* **300** (2003) 929–930, https://doi.org/10.1126/science.1169786.

L. J. Rothschild, R. L. Mancinelli, Life in extreme environments, *Nature* **409** (2001) 1092–1101, https://doi.org/10.1038/35059215.

M. J. Russell, A. J. Hall, D. Turner, *In vitro* growth of iron sulphide chimneys: possible culture chambers for origin-of-life experiments, *Terra Nova* **1** (1989) 238–241, https://doi.org/10.1111/j.1365-3121.1989.tb00364.x.

H. Stephen, Hydrothermal vents – life's first home?, SpaceRef, Ames Research Center, November 8, 2001, http://www.spaceref.com/news/viewpr.html?pid=6530.

K. O. Stetter, Hyperthermophiles in the history of life, *Phil. Trans. Roy. Soc. B* **361** (2006) 1837–1843, https://doi.org/10.1098/rstb.2006.1907.

D. Toomey, *Weird Life, The Seach for Life That Is Very, Very Different from Our Own*, W.W. Norton & Co., 2013, p. 131.

G. Wächtershäuser, Before enzymes and templates: theory of surface metabolism, *Microbiol. Mol. Biol. Rev.* **52** (1988) 452–484.

G. Wächtershäuser, Evolution of the first metabolic cycles, *Proc. Natl. Acad. Sci. USA* **87** (1990) 200-204.

G. Wächtershäuser, From volcanic origins of chemoautotrophic life to Bacteria, Archaea and Eukarya, *Phil. Trans. Roy. Soc. B* **361** (2006) 1787–1808, https://doi.org/10.1098/rstb.2006.1904.

Chapter 7

The Aromatic World, interview with P. Ehrenfreund, http://www.astrobio.net/interview/1992/the-aromatic-world.

M. Bernstein, Prebiotic materials from on and off the early Earth, *Phil. Trans. Roy. Soc. B* **361** (2006) 1689–1702, https://doi.org/10.1098/rstb.2006.1913.

R. Breslow, A likely posible origin of homochirality in amino acids and sugars on prebiotic earth, *Tetrahedon Lett.* **52** (2011) 2028–2032, https://doi.org/10.1016/j.tetlet.2010.08.094.

C. F. Chyba, J. J. Thomas, L. Brookshaw, C. Sagan, Cometary delivery of organic molecules to the early Earth, *Science* **249** (1990) 366–373, https://doi.org/ 10.1126/science.11538074.

C. Chyba, C. Sagan, Endogenous production, exogenous delivery and impact-shock synthesis of organic molecules: an inventory for the origins of life, *Nature* **355** (1992) 125–132, https://doi.org/10.1038/355125a0.

C. S. Cockell, The origin and emergence of life under impact bombardment, *Phil. Trans. Roy. Soc. B* **361** (2006) 1845–1856, https://doi.org/10.1098/rstb.2006.1908.

K. Harada, P. E. Hare, Analyses of amino acids from the Allende meteorite, In: *Biochemistry of Amino Acids*, Wiley, 1980, pp. 169–181.

Y. Kebukawa, A. L. D. Kilcoyne, G. D. Cody, Exploring the potential formation of organic solids in chondrites and comets through polymerization of interstellar formaldehyde, *Astronom. J.* **771**(19) (2013) 1–12, https://doi.org/10.1088/0004-637X/771/1/19.

K. A. Maher, D. J. Stevenson, Impact frustraton of the origin of life, *Nature* **331** (1988) 612–614, https://doi.org/10.1038/331612a0.

S. Pizzarello, Meteorites and the chemistry that preceded life's origin, *Prebiotic Evolution and Astrobiology* (J. Tze-Fei Wong and A. Lazcano, Eds.), Landes Bioscience, pp. 46–51.

S. Pizzarello, E. Shock, The organic composition of carbonaceous meteorites: the evolutionary story ahead of biochemistry, *Cold Spring Harb. Perspect. Biol.* **2** (2010) 1–19, https://doi.org/10.101/chperspect.a002105.

P. Schmitt-Kopplin, Z. Gabelica, R. D. Gougeon, A. Fekete, B. Kanawati, M. Harir, I. Gebefuegi, G. Eckel, N. Hertkorn, High molecular diversity of extraterrestrial organic matter in Murchison meteorite revealed 40 years after its fall, *Proc. Natl. Acad. Sci. USA* **107** (2010) 2763-2768, https://doi.org/10.1073/pnas.0912157107.

M. A. Sephton, Organic compounds in carbonaceous meteorites, *Nat. Prod. Rep.* **19** (2002) 292–311, https://doi.org/10.1039/b103775g.

N. H. Sleep, K. J. Zahnle, J. F. Kasting, H. J. Morowitz, Annihilation of ecosystems by large asteroid impacts on the early Earth, *Nature* **342** (1989) 139-142, https://doi.org/10.1038/342139a0.

C. Wickramasinghe, Life from space. Astrobiology and panspermia, *Features*, Evolution, The Biochemical Society, February 2009, 40-44.

Chapter 8

D. E. Canfield, M. T. Rosing, C. Bjerrum, Early anaerobic metabolisms, *Phil. Trans. Roy. Soc. B* **361** (2006) 1819–1836, https://doi.org/ 10.1098/rstb.2006.1906.

M. S. Dodd, D. Papineau, T. Grenne, J. F. Slack, M. Rittner, F. Pirajno, J. O'Neil, C. T. S. Little, Evidence for early life in Earth's oldest hydrothermal vent precipitates, *Nature* **543** (2017) 60–65, https://doi.org/10.1038/nature21377.

B. Fegley, R. G. Prinn, H. Hartman, G. H. Watkins, Chemical effects of large imacts on the Earth's primitive atmosphere, *Nature* **319** (1986) 305–308, https://doi.org/10.1038/319305a0.

J. F. Kasting, M. Tazewell Howard, Atmospheric composition and climate on the early Earth, *Proc Trans. Roy. Soc. B.* **361** (2006) 1733–1742, https://doi.org/10.1098/rstb.2006.1902.

R. Lill, Function and biogenesis of iron-sulfur proteins, *Nature* **460** (2009) 831–838, https://doi.org/10.1038/nature08301.

C. H. Lillig, R. Lill, Light on iron-sulfur clusters, *Chem. Biol.* **16** (2009) 1213–1214, https://doi.org/10.1039/b903811f.

W. F. Martin, Physiology, phylogeny, and the energetic roots of life, *Period. Biol.* **118** (2016) 343–352, https://doi.org/10.18054/pb.v118i4.4737.

J. Peretó, Controversies on the origin of life, *Int. Microbiol.* **8** (2005) 23–31, https://doi.org/10.1007/s0042500 50292.

V. D. Samuilov, Energy problems in life evolution, *Biochemistry* (Moscow), **70** (2) (2005) 246–250.

F. Westall, C. E. J. de Ronde, G. Southam, N. Grassineau, M. Colas, S. Cockell, H. Lammer, Implications of a 3.472-3.333 Gyr-old subaerial microbial mat from the Barberton geenstone belt, South Africa for the UV environmental conditions on the early Earth, *Phil. Trans. Roy. Soc. B* **361** (2006) 1857–1875, https://doi.org/10.1098/rstb.2006.1896.

Chapter 9

S. A. Benner, K. G. Devine, L. N. Matveeva, D. H. Powell, The missing organic molecules on Mars, *Proc. Natl. Acad. Sci. USA* **97** (2000) 2425–2430, https://doi.org/10.1073/pnas.040539497.

C. F. Chyba, Energy for microbial life on Europa, *Nature* **403** (2000) 381–382, https://doi.org/10.1038/35000281.

P. Ehrenfreund, M. Spaans, N. G. Holm, The evolution of organic matter in space, *Phil. Trans. Roy. Soc. A* **369** (2011) 538–554, https://doi.org/10.1098/rsta.2010.0231.

M. M. Grady, I. Wright, The carbon cycle on early Earth – and on Mars?, *Phil. Trans. Roy. Soc. B* **361** (2006) 1703–1713, https://doi.org/10.1098/rstb.2006.1898.

J. I. Lunine, Physical conditions on the early Earth, *Phil. Trans. Roy. Soc. B* **361** (2006) 1721–1731, https://doi.org/10.1098/rstb.2006.1900.

R. Papoular, The use of kerogen data in understanding the properties and evolution of interstellar carbonaceous dust, *Astronom. Astrophys.* **378** (2001) 597–607, https://doi.org/10.051/0004-6361:20011224.

C. Sagan, *The Cosmic Connection. An Extraterrestrial Perspective*, Hodder and Stoghton, Ltd., 1974.

D. Stevenson, Europa's ocean – the case strengthens, *Science* **289** (2000) 1305–1307, https://doi.org/10.1126/science.289.5483.1305.

P. Thaddeus, The prebiotic molecules observed in the interstellar gas, *Phil. Trans. Roy. Soc. B* **361** (2006) 1681–1687, https://doi.org/10.1098/rstb.2006.1897.

D. Toomey, *Weird Life. The Search of Life That Is Very, Very Different from Our Own*, W. W. Norton & Co., New York, 2014.

A. I. Tsapin, M. G. Goldfeld, G. D. McDonald, K. H. Nealson, B. Moskovitz, P. Solheid, K. M. Kemner, S. D. Kelly, K. A. Orlandini, Iron(VI): hypothetical candidate for the Martian oxidant, *Icarus* **147** (2000) 68–78, https://doi.org/10.1006/icar.2000.6437.

S. D. Vance, K. P. Hand, R. T. Pappalardo, Geophysical controls of chemical disequilibria in Europa, *Geophys. Res. Lett.* **43** (2016), https://doi.org/10.1002/2016GL068547.

ABOUT THE AUTHOR

Nenad Raos, born 1951 in Zagreb, Croatia, is a scientific adviser in chemistry, retired from the Institute for Medical Research and Occupational Health situated in his native town. He is the author of numerous books and articles on popular science and has spent seven years as the editor-in-chief of *Priroda*, the leading magazine for popularization of science in Croatia, continuously published since 1911. Dr. Raos is now the editor of the Chemical Education section in the journal *Kemija u industriji* (Chemistry in Industry).